THE ATLAS OF
NATURAL
WONDERS

THE ATLAS OF NATURAL WONDERS

RUPERT O. MATTHEWS

Facts On File Publications
New York • Oxford

A Marshall Edition
The Atlas of Natural Wonders
was conceived, edited and
designed by
Marshall Editions Limited
170 Piccadilly
London W1V 9DD

First published in the USA by
Facts on File Publications
460 Park Avenue South
New York, NY 10016

EDITOR   Pip Morgan
NATIONAL PARKS
AUTHOR   Anne Kilborn
MANAGING EDITOR   Ruth Binney
ART DIRECTOR   John Bigg
DESIGN ASSISTANT   Jonathan Bigg
PICTURE EDITOR   Zilda Tandy
PICTURE RESEARCH   Celia Dearing
PRODUCTION   Barry Baker
                    Janice Storr
                    Nikki Ingram

Originated by Alpha
Reprographics Ltd, Perivale, UK
Typeset by Hourds Typographica,
Stafford, UK
Printed and bound in Spain by
Printer Industria Gráfica S.A.,
Barcelona

Library of Congress
Cataloging-in-Publication Data

Matthews, Rupert.
   The atlas of natural wonders / Rupert O. Matthews.
      p.   cm.
   Bibliography: p.
   Includes index.
   ISBN 0-8160-1993-2
   1. Natural areas.   2. Natural monuments.   3. Natural history.
I. Title.
QH75.M38 1988
508--dc19
88-16387
   CIP

1 2 3 4 5 92 91 90 89 88

CONTENTS

INTRODUCTION

The surface of the Earth is a landscape with an ever-changing face. For many millions of years, the outermost skin of the globe has been sculpted continually into a wealth of wondrous forms. This malleable covering, caught between the planet's red hot core and the external forces of wind and weather, is punctured by volcanoes, raised up into formidable mountain ranges and etched by the forces of water and glaciers. The result is a staggering variety of spectacular phenomena.

This atlas charts the wonders that the powers of Nature have created, from the grand scale to the fine detail—from the birth of the Himalayas, thrust up from the sea bed, to the intricate coral gardens of the Great Barrier Reef; and from the Amazon's mighty waterways to the finely wrought fairytale décor in Frasassi's limestone caverns. By blending their geography and history with the story of their geological past, each natural wonder is revealed as unique. And each has a singularity that deserves preservation from the ravages of mankind.

Yet even if we manage to protect our planetary heritage for the next generation, the natural wonders of today do not have an infinite lifespan. Alterations in climate, the slow drift of continents and the continuous processes of erosion mean that our superlative landscapes will inevitably be diminished in the fullness of geological time. But gradually they will be replaced by fresh wonders, whose dimensions and locations can only be guessed at.

The world map shows the locations of the natural wonders featured in this atlas. The book has been organized on the basis of longitudes—starting at the Greenwich Meridian and working eastward. Places in bold type are the subjects of major essays; the remainder are described in the Gazetteer section on pages 218–223. The longitude and latitude of each of the natural wonders have been given to one decimal place.

Strokkur

Sognefjord

Lake Vänern

Skye
Giant's
Causeway
Ben Bulbin

Waddenzee

Cheddar Gorge
Chesil Bank

Königssee
Mont Blanc
Eisriesenwelt Caves
Armand Cave
Camargue
Lake
Maggiore
Plitvice Lakes
Gorge du Verdon
Frasassi Caves
Mount Vesuvius
Metéora
Mount Etna
Pamukkale Springs
Ürgüp Cones

Las Marismas

Nefta Oasis

Dead Sea
Negev Desert

Great Western Erg

Ahaggar
Mountains

Lake Chad

Congo
Basin
Ruwenzori
Victoria Nyanza
Ngorongoro Crater
Mount Kilimanjaro

Victoria Falls

Madagascar

Namib Desert

Table Mountain

Volga

Caspian Sea

Lake Baikal

Gobi Desert

Huang
He

Vale
of Kashmir
Band-e
Amir Lakes
Himalayas
Mount Everest
Ganges
Western Ghats
Mekong

Nile
Red Sea

Matsushima Bay

Mount Fuji
Beppu

Guilin Hills

Mount Mayon

Mariana Trench

Lake Toba

Krakatoa

Great Barrier Reef

Ayers Rock
Simpson
Desert

Whakarewarewa

Milford
Sound

Ross
Ice Shelf

80°
60°
40°
20°
0°
20°
40°
60°

20° 0° 20° 40° 60° 80° 100° 120° 140° 160°

7

GREAT WESTERN ERG

Ever-shifting sands of the Sahara Desert

The road leading from the Saharan Atlas Mountains to the Algerian town of Laghouat offers a rare vantage point from which to view the Great Western Erg. Vast hills of sand stretch out to the horizon like an ever-shifting carpet. Smooth-sided, sharply-crested dunes, each bewilderingly alike, imitate the succeeding waves of a sea.

The dune fields of the Great Western and other ergs in the desert constitute only about one fifth of the Sahara's total area of around 9 million sq.km (3.5 million sq.mi). The bulk of the world's largest desert is composed of monotonous plains of gravel, sterile rocky plateaus, arid mountain peaks and mirage-filled salt flats.

The endless procession of the erg's lazy, shifting dunes is both awesome and disorientating. Gazing at the barren but majestic land, it would be difficult to argue with the Algerian novelist Albert Camus (1913–1960) who described the desert as 'a land of useless and irreplaceable beauty'. The ever-blowing wind sculpts the sand into fantastic shapes only to destroy them again. The sun beats down with merciless intensity, making this one of the hottest, driest places on Earth. The burning temperature of the sand causes a quivering of the air, blurring the edges of the landscape and confusing the eye.

The top layer of sand on the surface of the erg (the word comes from the Arabic, meaning 'large area of sand') is subject to the caprices of the wind. Where strong steady winds blow from a single direction, this sand is heaped up to form great crescent-shaped dunes, or barchans, that edge their way across the desert. In confused air patterns, where swirls and eddies mean a constantly changing wind direction, the dunes are piled up into a complex assortment of shapes. In some areas they form

The Great Western Erg, or Erg Occidental, occupies the northwestern region of the Sahara Desert (*left*). It covers an area of 78,000sq.km (30,000sq.mi), roughly the size of the state of South Carolina in the USA. Huge, crescent-shaped sand dunes, or barchans, drift across the desert (*right*). They are rolled forward by steady prevailing winds and often travel 30m (100ft) in a single year.

long parallel lines separated by broad troughs. Elsewhere, converging air currents stack the sand up into huge mounds or pyramids which can reach 120m (394ft) in height.

Around 15,000 years ago, when most of North America and Europe were in the grip of an ice age, the Great Western Erg – and the rest of the Sahara – were rich and fertile. Herds of animals roamed the rolling plains and lush grasslands. Prehistoric man lived and thrived here, leaving behind cave paintings which depicted the land as a hunters' paradise.

Gradually the climate changed across the globe. The glaciers melted and receded. Moisture-laden air currents from the Atlantic Ocean moved north to precipitate their water over Europe instead of Africa. Now dominated by dry winds, the Sahara was deprived of its streams and rivers. With no moisture to bind it together, the soil broke down and lost its fertility. In a comparatively short span of geological time, the rich earth became a sea of shifting sand. Today, the rains come hardly at all and, when they do, they last only a few hours.

Study of the long-term weather patterns of the Earth suggests the Sahara will not always remain a desert. Sooner or later it will experience another wet and fertile time. Over the past 3 million years the region has fluctuated between wet and dry periods, just as northern Europe has undergone a cycle of ice ages. The sand in the Great Western Erg was left over from the previous wet period when rivers and streams washed huge quantities of sand, soil and sediment from the mountains to the north.

Of ghouls and sandstorms

Strange cries have been heard at night, furtive figures seen scurrying through the shimmering dunes or following in the footsteps of camel caravans. Such visions have given birth to tales of ghouls. Nomads swear they have seen these terrible man-eating fiends, pointing as evidence to the bones in the desert and asking: 'What other than ghouls could pick bones clean in the desert?' The answer is probably sand. And sand, whipped by wind into a small column, could easily be mistaken for a supernatural being when viewed through the quivering air.

The wind may rush across the Great Western Erg in frantic fury and unslowed by obstacles. Large quantities of sand are swept up in its wake to form a blinding cloud. A 19th-century British traveller, Sir Samuel White Baker (1821–1893), described the onrush of a sandstorm: 'I saw approaching from the southwest apparently, a solid range of immense brown mountains, high in the air. So rapid was the approach of this extraordinary phenomenon, that in a few minutes we were in actual pitchy darkness . . . We tried to distinguish our hands placed close before our eyes, not even an outline could be seen.'

Such sandstorms often travel at around 48km (30mi) an hour. The solid wall of swirling sand that appears almost without warning can reach 1.6km (1mi) in height and present a front up to 480km (300mi) wide. Battering winds muffle all sound, sand blocks out all light.

A series of definite layers forms within a sandstorm. From the ground to about waist high, the air is thick with gravel and coarse sand. Above is a layer of thinner sand that penetrates every crack and crevice. Only the finest sand and dust is whirled high into the air – often enough to block out the sun many miles from the heart of the storm. And the effects of such storms are not always restricted to the desert. Saharan dust is occasionally carried across the Mediterranean Sea into southern Europe and even to Britain. After a storm in Algeria in 1947 parts of the Swiss Alps were turned pink with red Saharan dust.

Camel caravans have graced the North African deserts since the 3rd century AD, when dromedaries were introduced from Arabia. Camels, in fact, originated in North America but the native types disappeared from that continent more than 10,000 years ago.

The Arabian camel (*Camelus dromedarius*) is one of the best-adapted desert mammals. Its broad, two-toed, thick-soled feet prevent it from sinking into treacherous desert sands while the fat store in its single hump can easily be converted to a source of water in times of drought. An insulating layer of air within its hairy coat protects the creature from overheating in the burning sun.

The onslaught of sand and dust in desert storms has little effect on a camel's performance. Long, bushy eyelashes and thick lids protect its eyes; special muscles in its face allow the camel to open and close its nostrils (*above, right*) at will; the camel's hairy coat softens the impact of the potentially harmful effects of flying sand.

Crescent-shaped sand dunes, or barchans, are formed in dry deserts where strong, steady winds blow predominantly from a single direction. Sand grains blown by the wind over a barchan's crest double back on themselves to form a steep,

leeward face (*right*). Because this is a constant process, the barchan moves across the desert at speeds of up to 30m (100ft) per year. In wide, open areas, barchans may congregate in large, triangular formations (*below*) that resemble flocks of geese.

HOW A BARCHAN FORMS

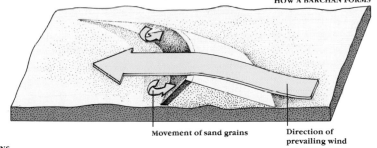

Movement of sand grains   Direction of prevailing wind

GROUPS OF BARCHANS

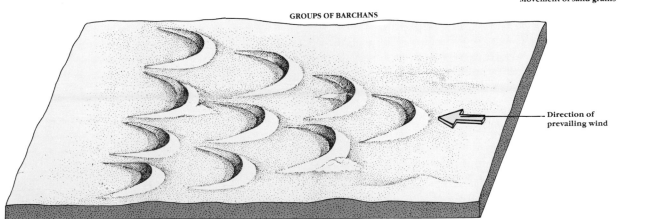

Direction of prevailing wind

11

SOGNEFJORD

Sculpted legacy of the ice ages

The fjords that shape Norway's western coast are a spectacular monument to the awesome power of glaciers. The undisputed king of them all is Sognefjord: in fact, no other fjord in the world is as long, as deep or as majestic. Sheer cliffs and snowcapped mountains rise from the water's edge; picturesque villages and farms nestle in every habitable area of land. Waterfalls cascade through forested slopes or over bare rock into the waveless waters of the fjord.

From the Solund Islands on the edge of the North Sea, Sognefjord extends inland for a distance of 200km (125mi). Its sheer cliffs rise almost vertically, in places to a height of 900m (3,000ft). At one point, near the town of Vadheim almost a third of the way in from the sea, Sognefjord descends to a depth of 1,234m (4,078ft). From the summit of its cliffs to the bed of its waters, the total depth of Sognefjord's chasm is a third as much again as the gorge of Grand Canyon in the USA.

For much of the Earth's lifespan of approximately 4,600 million years, its climate has been hot and dry, its continents devoid of ice. But geologists have detected seven ice eras, each lasting around 50 million years, in which cooler, wetter climates have predominated. The most recent of these ice eras began around 65 million years ago and has been marked by six ice epochs, when global temperatures bring especially cold weather. Each epoch spans roughly 2.5 million years.

Within the epochs are periods, known as ice ages, of exceptional cold and these are coupled with extensive glaciation. The last ice age started around 125,000 years ago, peaked some 50,000 years ago and ended in about 8,000 BC. For most of this ice age, temperatures at the poles were far colder than they are today. Huge ice sheets covered much of northern Europe and

Sognefjord forms a huge inlet on the southwestern coast of Norway, around 240km (150mi) northwest of Oslo, the country's capital, and 80km (50mi) north of Bergen (*left*). The headwaters of Sognefjord's easternmost branch, Årdalfjord, flow over the Vettis Falls which, with a drop of 275m (900ft), are the highest in Norway. Around Leikanger (*right*), situated roughly halfway along the fjord's northern shore, fruit trees thrive in average annual temperatures of 13.2°C (55.8°F).

North America. Flora and fauna from these regions retreated southward. The habitats on the threshold of the ice sheets were tundra and bleak scrubland, populated only by mammoths, hairy mastodons, woolly rhinoceroses and musk oxen.

In each successive ice age, Scandinavia was submerged under a gigantic ice sheet for thousands of years at a time. At Sognefjord, an ancient river system was covered by glaciers that moved down from the surrounding mountains. As glaciers appeared and reappeared, they gouged out more and more rock and earth from the side valleys as well as from the main river valley itself.

The glacier which, during the last ice age, created Sognefjord's present shape was thickest at the head of the fjord, around 210km (131mi) from the open sea; estimates put its thickness here at 3,000m (9,840ft). About 50km (31mi) from the sea, the glacier was at its thinnest – between 100m (330ft) and 300m (1,000ft) thick. When the glacier finally receded around 10,000 years ago, the waters of the North Sea flooded in and filled a classic glaciated valley: U-shaped, flat-bottomed, its rock walls scratched with lines, and less deep at the seaward end where the glacier lost its power of erosion.

From sea to snowcapped mountains

The dramatic scenery of Sognefjord can best be seen from the ferries that daily ply its waters. Beside the mouth of the fjord, bare rounded hills rise steeply from the sea. Covered only by the thinnest soil with limited fertility, these low hills can support only a small population. Farther inland, the sheer cliffs rise so steeply from beneath the water that large ferries can pass to within an arm's reach of them without running aground.

At the town of Balestrand on the north bank, the fjord of Fjaerlands branches off for 25km (15mi). Its waters are fed by the Jostedal Glacier, Europe's largest ice sheet and a remnant of the glaciers that created Sognefjord. After the ice age ended, the Jostedal Glacier is thought to have disappeared altogether. But it returned during the 'little ice age', between the 15th and 19th centuries, and attained an estimated thickness of 300m (1,000ft).

After bending northward between Balestrand and the ancient Viking town of Vik on the south bank, Sognefjord begins to branch into smaller, tributary fjords where the climate is warmer and wetter. These are rarely deeper than 300m (1,000ft), whereas the water in the main trunk of Sognefjord is seldom shallower than 800m (2,625ft).

At the point where Sogndalfjord branches off to the north and Aurlandsfjord to the south, an increasing amount of land is used for farming and fruit-growing purposes. The town of Liekanger alone has more than 60,000 fruit trees, mainly peaches, apricots and walnuts. In these branching fjords, and at the head of Sognefjord, the land has risen around 100m (330ft) since the end of the ice age. This elevation was caused by the inrushing sea, which brought vast amounts of sand and gravel to form delta fans and narrow terraces at the foot of the retreating glaciers.

Behind the town of Kaupanger on the north bank, Norway's highest continuous slope rises more than 915m (3,000ft). After this point, Sognefjord terminates in three branches: Lærdalfjord, opposite Kaupanger to the south; Årdalfjord to the east; and Lusterfjord to the north.

Lusterfjord, Sognefjord's longest branch, and one of its broadest, extends for 48km (30mi). At its tip is Sognesfell, a pass which leads to the Jotunheim, which means 'home of the giants' in Norwegian. Geologically related to the ancient rocks of Scotland, these mountains contain the highest peak in Scandinavia – Glittertinden, which rises to 2,472m (8,110ft).

Elaborately carved limestone memorials were erected by the Vikings in honour of their dead heroes. This picture stone (*below*), which stands 3.7m (12ft) high, is one of many memorials on Sweden's island of Gotland. The four panels depict, from the top: the hero's death in battle; his funeral; his entrance into Valhalla, the warrior's heaven; and a longship, symbolizing the passage of the soul.

Prows of Viking longships were always adorned with fierce-looking figureheads, such as this 9th-century stem post, carved from oak into a dragon's head (*below*). Such wooden sculptures were not fixed to the ships' bows for purely ornamental reasons; they were deliberately installed to spread fear into the hearts of Viking enemies and to ward off evil spirits during their long sea journeys.

The reindeer (*Rangifer tarandus*) migrates in winter to parts of southern Scandinavia, such as Sognefjord, where it feeds on lichens growing beneath the snow. In summer, reindeer return to the tundra regions of the north to breed. Both sexes grow antlers, but those of the male (*below*) have larger branches.

The small town of Vik stands at the edge of a saucer-shaped bay, midway along Sognefjord's southern flank. Ringed by soaring, snowcapped peaks, Vik (which means 'bay') is renowned for being Norway's largest producer of the 'old cheese', or gammelost.

The town's history reaches back to the Viking era (800–1050 AD), when Norsemen set out from their farms on their voyages to various parts of the world.

Generations of skilled craftsmen applied the same techniques and designs to both Viking ships and church architecture. Vik's 12th-century, pagoda-roofed church of Hopperstad, for example, is decorated with winged serpents and other beasts, fashioned from wood like the stem posts of longships. The carved timber of the choir screen and one of the altars are reminiscent of ships' beams.

GORGE DU VERDON

Canyon forged in a limestone landscape

Awesome in their splendour, the steep cliffs, plunging canyons, and cascading torrents of Verdon are hidden deep in the mountains of southeastern France. At the heart of this limestone landscape, the turbulent waters of the River Verdon have carved a gorge so spectacular that it has become known as the 'Grand Canyon of Europe'.

In 1905, the French pot-holer and speleologist E.A. Martel became the first man to explore the length of the Verdon Gorge and to draw the world's attention to its grandeur. The canyon runs for 20km (13mi) between the medieval towns of Castellane to the east and Moustiers-Ste-Marie to the west. On each side of the Verdon's rushing waters, precipitous cliffs rise to more than 700m (2,300ft) in height. At the waters' edge, where bare stone walls flank the river, the gorge narrows to as little as 6m (20ft), then widens again to 100m (330ft). Above the line of the winter floods, the naked limestone gives way to scrub and bushes. At the top of the gorge, the span varies between 200m (660ft) and 1.6km (1mi).

The landscape of the limestone gorge and its surroundings was not created entirely by the power of the Verdon's torrent. The small amounts of carbon dioxide gas that rain collects as it falls through the air turns it into carbonic acid. This weak acid penetrates and widens even the smallest crack by dissolving and washing away calcium carbonate, the chief constituent of limestone. As a result, the rains falling on the limestone massif around the River Verdon have slowly eaten away the rock, creating a multitude of caves which open out on to the walls of the gorges.

This honeycombing of the limestone weakens the whole rock area and makes it easier for the river to cut its downward path. Geologists have suggested that, before the Verdon Gorge was

The Verdon Gorge is situated in the limestone massif of Provence around 128km (80mi) northwest of Cannes in the southwest corner of France (*left*). The River Verdon, which carved the gorge, rises at Mount Pelat, a peak of some 3,000m (10,000ft) in the Alpes Maritimes. It flows 190km (120mi) to join the River Durance – a tributary of the Rhône – around 50km (31mi) north of Marseilles. The Verdon Gorge (*right*), the longest in France, is the most spectacular stretch of the river's series of limestone gorges.

formed, the river flowed through an underground cavern. As chemical erosion slowly weakened the roof of this cavern, there came a point where the roof collapsed under its own weight and opened up a great chasm. The river's torrent would have washed away the massive volume of debris generated by such a fall, leaving behind the gorge as it stands today.

In the vicinity of the gorge

Before it reaches the spectacular series of gorges, the river flows southward into the town of Castellane from a clear blue lake: around 9.6km (6mi) long, it was created by the Castillon Dam for hydroelectric purposes in 1947. In the town itself, the waters hurry beneath Le Roc, a solitary pinnacle around 183m (600ft) high and capped by a tiny chapel dedicated to the Virgin Mary.

When viewed from the top of Le Roc, the river disappears between the perpendicular walls of the gorges' narrow entrance. A road offering breathtaking views from a number of vantage points has been built around the Verdon Gorge. Les Balcons de la Mescla, for example, is a natural balcony on the southern wall overlooking the *mescla*, or 'mixing', of the Verdon's waters with those of a tributary, the Artuby. The rock walls opposite the balcony reveal a series of granite ledges variously tinted in shades of red or yellow ochre.

Beneath the balcony, the Verdon abruptly changes course toward the northwest from its previous southward direction. The Corniche Sublime, a mountainous road built high up on the south side of the canyon in 1947, winds in and out of tunnels, and brings travellers to further vantage points, such as Les Falaises des Cavaliers and Le Col d'Illoire.

At the western end of the gorge, the river flows into the artificial lake of Sainte-Croix. Created in 1973, at the same time as the Rue des Crêtes along the north wall of the gorge, the lake's 2,500 hectares (6,667 acres) are used primarily for recreational purposes. To the north of the lake stands the town of Moustiers-Ste-Marie, said to have the most beautiful setting in all Provence.

The river's green waters

Tall rocks stand up from the centre of the river bed, stirring the gorge's torrent into long streamers of white water, which contrast sharply with the Verdon's delicate shade of green. In the full glare of sunlight, the river reflects an almost turquoise colour. But the waters, while free of soil and organic debris, are not clear. In fact, as the Verdon flows through the gorge, it has an almost milky appearance.

The green translucence of the river originates at the Verdon's source at Mount Pelat. The snowfields and glaciers of the Alpes Maritimes erode the underlying rock by a combination of direct pressure and below-freezing temperatures. The resulting pulverized rock is washed into the Verdon where, as fine particles, it is held in suspension. The physical interplay between the sun's rays and the fine particles conspires to reflect only the green-blue part of the visible spectrum.

More than 2,000 years ago, the unusual colour of the Verdon inspired a religious cult among the Vocontii, the dominant tribe of Ligurian Celts who ruled the region. Their pagan religion placed great emphasis on the power of nature gods and spirits, especially those associated with unusual natural features. River goddesses, to whom the Ligurian Celts frequently offered sacrifices, were among the most powerful deities. The Vocontii worshipped the Goddess of the Green Waters and were reputed to hurl votive offerings into the River Verdon to appease the goddess or to seek her favour.

Behind the battlemented tower in the ancient Provencal town of Castellane rises the sheer cliff face known simply as Le Roc. The view from the little 17th-century Chapelle Notre-Dame du Roc on top includes the point where the River Verdon vanishes into the deeply-cleft entrance to the Verdon Gorge.

The gorge's winding progress can easily be followed from high above the south bank of the river, travelling along the spectacular Corniche Sublime.

The medieval town of Moustiers-Ste-Marie, noted as the centre of a centuries-old *faience* pottery industry, rests at the western end of the sequence of gorges, north of the artificial lake of Sainte-Croix into which the River Verdon empties.

The Moustiers or monasterium from which the town takes its name was founded here in 432 by monks from the neighbouring town of Riez. Moustiers straddles the pocket between two vast cliffs, swarming part way up the sides of each, and basking in the shade of cypress trees.

MONT BLANC AND THE ALPS

Snow-white summit of a mountain chain

On a summer's day in 1946, the foreman of a northern Italian wool factory received a puzzling phone call from his millionaire employer, Count Dino Totino. 'Stop the looms,' the nobleman cried excitedly. 'Send every man to join me at Mont Blanc. We are going to dig a tunnel through to France.' Not daring to disobey such a direct command, the foreman packed his entire work-force off to the Alps. Though they toiled hard, the textile workers tunnelled only a distance of 50m (160ft) before winter forced them to down tools.

In 218 BC, the Carthaginian general Hannibal, together with his army of elephants and 40,000 men, took 15 days to cross the Alps. Even after World War II, the only route around Mont Blanc took 18 hours to traverse and was impassable for half the year. Such a marathon journey, together with the realization that a tunnel would promote trade and friendship with France, prompted Count Totino to initiate his scheme. But the Italian and French authorities each suspected that the other would use the tunnel as a point of invasion. As a result, Totino was ordered to stop digging. In 1959, however, he finally persuaded the Italian government to support his plan. Six years and 17 deaths later, the tunnel through the solid rock of Mont Blanc was complete.

The mountain range that is the Alps runs in an arc for around 1,200km (750mi) through seven countries: France, Switzerland, West Germany, Italy, Lichtenstein, Austria and Yugoslavia. At its broadest, the range is roughly 300km (187mi) across. Of all the alpine peaks and summits of Western Europe, Mont Blanc, the 'white mountain', is the highest at 4,807m (15,771ft).

The mighty forces that moulded the spectacular Mont Blanc were also responsible for creating the entire alpine chain. Around 40 million years ago, two sections, or tectonic plates, of

Mont Blanc is the highest peak on a huge alpine massif that overlooks the Chamonix valley in eastern France, where the country borders with Italy and Switzerland (*left*). Scarred by glaciers and eroded into many *aiguilles*, or needles, the Mont Blanc massif bristles with 10 peaks more than 4,000m (13,123ft) high. To the east of Mont Blanc's peak, the Aiguille Verte towers over the still waters of Le Lac Blanc (*right*).

the Earth's crust started to drift toward each other. One, bearing the continent of Africa, gradually moved north and collided with the plate bearing Europe. As the two continents ground inexorably together, the rocks between them buckled, folded and were finally forced upward.

The bulk of these rocks were sedimentary, formed on the beds of an ancient sea. Yet tough granite and schist were also caught in the upheaval. An amalgam of these extremely hard rocks form the rugged backbone of the Alps. They have largely resisted the natural forces of erosion and form many of the great peaks in the mountain range, including Mont Blanc and the Matterhorn. At the margins of the Alps, the softer, sedimentary rocks, such as limestone and sandstone, have been weathered into a smoother landscape. During the last ice age, which terminated around 10,000 years ago, glaciers gouged out U-shaped, flat-bottomed valleys from many mountainsides, and so fashioned some of the most spectacular scenery in the world.

The difficulties of scraping a living from the thin soils of the high alpine slopes led the ancient inhabitants to adopt a seasonal lifestyle. During the Middle Ages, the village farming community dwelt in wooden, flat-roofed houses and planted crops in the fertile valley soil in spring. Early in summer, they drove their herds of cattle and sheep to the mountains and left them to graze on the rich grass that sprouted up after the snows melted. Though declining in many areas, this alpine way of life still thrives in some remote districts.

The changing face of the Alps
Since the early 19th century, the alpine inhabitants have undergone a major change in lifestyle. At first, visitors came to the region for climbing or walking holidays, which baffled the locals who could not understand why people would wish to climb mountains or trek long distances for amusement. But before long they discovered they could profit from these visitors by acting as guides. After World War II, hiking and mountaineering gave way to the sport of skiing. Once again the alpine folk changed their ways to capitalize on the people who were turning the mountains into a playground.

However, the popularity of the Alps may be their ruin. Skiing resorts, which cater for huge influxes of visitors, demand heavy development of the mountains. In the 1980s, as many as 50 million people visit the Alps each year. More than 40,000 ski runs criss-cross the slopes. In Austria alone, which contains 35 per cent of the Alps, there are ski runs with a total length of 6,000km (3,745mi) – 200km (125mi) longer than the country's entire railway network.

Long strips of alpine forest are bulldozed to provide skiers with the wide, smooth pistes they seek. As a result, the mountain slopes are destabilized. Tree roots bind the soil together, preventing erosion and gully formation. Branches hold snow off the ground and trunks prevent drifting. Without trees to hold the soil and snow, the pistes become ideal paths for avalanches and mudslides on a grand scale. In less than 20 days during the summer of 1987, mudslides killed 60 people and damaged nearly 50 towns and villages in the alpine region.

Smoke and fumes from many industrialized areas of Europe find their way to the mountains to attack the forests on another front. Chemical pollutants in these emissions are affecting as many as two-thirds of the trees in some areas, while Bavaria has reported that 78 per cent of its forests have been damaged beyond repair. Trees weakened by pollution become bare or frail and, as such, are more prone to disease or gale damage.

At the birthplace of mountaineering, on Mont Blanc, a stone crucifix stands as a memorial to fallen climbers. In a small cemetery near the Mer de Glace station rest some of the world's foremost mountaineers. Among them are Edward Whymper, the conqueror of the Matterhorn, and Louis Lacheral, the first man to climb the Himalayan peak of Annapurna.

English mountaineer Edward Whymper
(1840–1911) (*right*) led the expedition which, on July 14, 1865, conquered the Matterhorn. This pyramidal peak stands 4,505m (14,782ft) high on the Swiss-Italian border to the southwest of Zermatt. The triumphant ascent was marred by the tragic deaths of Michel Croz (*left*), reputed to be the finest alpine guide of his generation, and three British mountaineers – Lord Francis Douglas, the Reverend Charles Hudson and Douglas Hadow.

The striking panorama seen looking northeastward from the summit of Mont Blanc (*above*) features the jagged peaks of the massif on the French-Italian border. In mid-distance, the spiky Diable Arête descends from Mont Blanc du Tacul into the white ice field of the Géant Glacier. Beyond stands the Aiguille Verte (on the left of the picture), one of the many rocky needles on the massif. Behind the Grandes Jorasses (on the right of the picture) the peaks of the Swiss Pennine Alps peep through the clouds: (from left to right) the Weisshorn, Grand Combin, Matterhorn and Monte Rosa.

NEFTA OASIS

Green garden in a desert wilderness

The first spring of freshwater to issue from the earth after the flood was discovered, according to legend, at Nefta by Kostel, one of Noah's grandsons. From that spring developed a fertile oasis which today nestles among barren, rolling hills in the west of Tunisia. A startling patch of green in a sea of sand, Nefta overlooks on its southern side a vast saline depression known as the Chott Djerid.

While the blazing desert sun bakes the surrounding land, sweet water wells up from beneath Nefta's soil. The precious liquid emerges through 152 springs which are fed, not by local rain, which is negligible, but from water falling on distant hills. This percolates down through the sand until it reaches a layer of porous rock, then seeps horizontally underneath the desert.

Where the water-laden, porous rock lies close to the desert's surface, as at Nefta, springs known as artesian wells puncture the sand and irrigate the land. In all, more than 950 hectares (2,347 acres) of desert have been made fertile by Nefta's artesian wells. In the 1960s, these natural springs were complemented by a number of manmade outlets.

Immediately to the north of the oasis lies a palm-embroidered hollow named La Corbeille, which in French means 'the open basket'. Here, the spring water issues forth from dozens of cracks and fissures in the steep brown slopes of this crater-like depression. In his book *Fountains in the Sand* (1912), British travel writer Norman Douglas (1868–1952) described La Corbeille as 'a circular vale of immoderate plant-luxuriance, a never-ending delight to the eye'.

When Douglas viewed the multitude of springs emanating from the walls of sand he saw them forming 'glad pools of blue and green that mirror the foliage with impeccable truthfulness

Nefta Oasis lies among desert hills and dunes, some 400km (250mi) southwest of Tunis, at the northwest corner of the saline depression known as Chott Djerid (*left*). A fertile way station for desert caravans, and a religious centre for Sufi pilgrims, Nefta provides a stopping-off point for birds migrating to and from the continent of Africa. Fed by more than 150 springs, the oasis features date palms as its predominant vegetation (*right*).

and then, after coursing in distracted filaments about the "corbeille", join their water and speed downhill towards the oasis, a narrow belt of trees running along either side'.

Of Sufis and dates

For more than a thousand years, Nefta has been an important religious centre for a mystical Islamic sect known as Sufis. The founder of this centre, Sidi Ibrahim, arrived at Nefta soon after Moslems conquered the region in AD 670. His aim was to seek the solitude he needed to study Islam's holy book, the Koran, and to contemplate the will of Allah.

Sufi teachings maintain that there is a hidden meaning in the words of the Koran, which can only be disentangled during patient study. Because this entails looking deeply into the holy book and not accepting it at face value, Sufis often came into conflict with orthodox Islam. In the 11th century, a compromise was reached between the two when Sufism was regarded purely as a way of 'apprehending reality', not as a means of interpreting the words of the Koran.

Tunisian Sufis established zaouias, or religious fraternities, throughout the country, especially in rural areas, where local people were offered shelter, education and justice for their grievances. At Nefta, the Zaouia of Sidi Ibrahim overlooks La Corbeille and contains not only the tomb of the saint, but also those of his son and numerous disciples. The holy reputation of the oasis brought other men of religion, such as Sidi Bou Ali, who enhanced still further the spiritual nature of the place. As a result, 24 mosques and more than 100 shrines stand in the old quarters of Nefta, their white cupolas shimmering in the sun.

An intricate network of seguias, or irrigation ditches, distributes spring water to nourish Nefta's fertile gardens. Orderly rows of date palms, some 350,000 trees in all, grow in square plots of land. Around one third of the trees belong to the Deglas variety, reputed to be the finest dates in the world. The delicious, energy-rich dates, the so-called 'fingers of light and honey', are not the only product of the palm. Leaves are cut into strips which, after drying, are used to weave baskets and other containers. When the palms stop producing, which may be after 200 years, their wood is used as timber. Date stones are ground up and fed to livestock, while the sap is made into a wine known as 'lagmi'.

Treacherous mud flats

To the south of the oasis stretches the notorious Chott Djerid, a vast seasonal lake which alternates from water through mud to parched plain every year. In autumn, the water table underlying Nefta's artesian wells rises dramatically and spills out to flood the Chott. In spring, the water table drops and the lake becomes an immense swamp of salty mud. The summer sun bakes the surface to a hard crust, creating the illusion of a firm plain, yet underneath the ground is soft. In the 1980s, the Tunisians built a highway across the Chott, from Tozeur – some 24km (15mi) east of Nefta – to Kebili, some 90km (56mi) to the southeast.

In the past, a safe path of solid ground was waymarked across the Chott with trunks of palms. Unwary travellers who strayed from the prescribed route would perish in the treacherous mud. A 12th-century Arab writer reported how a caravan of 1,000 camels was swallowed up by the desolate wastes. 'Unfortunately one of the beasts strayed from the path, and all the others followed it. Nothing in the world could be swifter than the manner in which the crust yielded and engulfed them; then it became like what it was before, as if the thousand baggage camels had never existed.'

The rows of sturdy date palms at Nefta Oasis offer protection against the heat of the sun, which can reach 45°C (113°F) in the shade in summer, and against desert winds, such as the scirocco. Beneath the palms, Nefta's inhabitants grow fruit trees, especially figs, pomegranates, apricots, peaches, oranges and lemons. The soil under these trees is further used to grow vegetables for the home or for the local market.

The date palm (*Phoenix dactylifera*) produces heavy clusters of brown fruits which are sweet and exceptionally nutritious. One of the world's oldest cultivated plants, the date palm grows up to 30m (100ft) high and may live for as long as 200 years. Trees reach maturity at the age of around 15 years, and thereafter produce an annual harvest of as much as 90kg (200lbs) each.

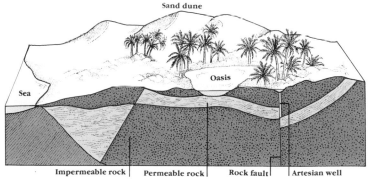

Nefta and other oases owe
their fertility to spring water
which emerges from artesian
wells or collects in pools formed
when the water table reaches the
surface of the desert. Rain falling
some distance away from the
oasis, such as beside the sea,
percolates horizontally through
layers of permeable rock, a
process that may take many years.

Artesian wells are commonly
located along fault lines in the
Earth's crust where layers of rock
have cracked and shifted. The
weight of the accumulated water
in the permeable rock layer
pressurizes the water in a well
and forces it upward through the
fault line.

LAKE VÄNERN

The heart of Sweden's waterways

The beautiful, forest-fringed waters of Lake Vänern nestle in Sweden's agricultural heartland. This enormous expanse of water, the home of rich and abundant wildlife, has long been the focus of human activity. More than 5,000 years ago, the ancestors of the Goths founded the Swedish civilization here. From a plateau between the twin peaks of Halleberg and Hunneberg that overlook the lake's southern shore, warriors of old crossed the 'rainbow bridge' to Valhalla, the mythological resting-place of Scandinavian heroes.

Lake Vänern fills part of a broad basin created more than 500 million years ago when powerful movements of the Earth's crust forced the region's hard volcanic rocks to subside. The lake covers an area of around 5,546sq.km (2,170sq.mi), which makes it the largest body of freshwater in Western Europe. Its waters reach a depth of 98m (320ft), although they are often substantially shallower than this. A ridge of hills bisects the basin, appearing on the lake as the north-south peninsulas of Värmlandsnäs and Kallåndshalvö, and an archipelago of islands.

Low hills and mountains cradle the lake to the south and east: on one side the river and streams flow into Vänern, while on the other they flow to the Baltic Sea. To the north, rolling hills rise toward the mountainous spine dividing Sweden from Norway; here the large rivers that feed Vänern, such as the Klarälv, have their origins.

The lake's only outlet is to the south along the Götaälv River. The waters of the river have cut a path through a ridge of gneiss rock, an ancient rock hardened by intense temperatures and pressures. As it slopes away to the south this ridge drops more than 30m (100ft) in just under 1.6km (1mi). Until the waters were diverted into a hydroelectric scheme during the first half of

Lake Vänern lies in southern Sweden, some 75km (47mi) north of Gothenburg and more than 200km (125mi) west of Stockholm, the country's capital (*left*). The vast lake, which measures around 130km (81mi) in length with a maximum width of some 75km (47mi), is largely fed by rivers flowing into its northern waters. Each of the islands and islets (*right*) is growing in area, since the water level of the lake falls by about 8cm (3in) per century.

the 20th century, Vänern emptied itself over the slope in the form of the impressive Trollhättan Falls. On a visit to the falls at the end of the 19th century, English artist Charles Lovett declared that: 'As the wild waters rush by with tremendous uproar, one has the feeling that no power short of Omnipotence could stay them in their mad onward rush.' In the 1980s, the falls could be seen in their former glory every July on Waterfall Day when the power station was temporarily closed down.

The Göta Canal

In 1718, King Charles XII of Sweden commissioned engineer Christopher Polhem to build a series of eight locks around the falls so that boats could travel between Lake Vänern and Gothenburg on the North Sea coast. But work on the project was halted in the same year when the king died on the field of battle. Despite fresh channelling attempts in 1749, the Trollhättan Canal was not completed until 1800. Nine years later, entrepeneur Baron Baltzar von Platen convinced the Swedish Parliament to sponsor a canal link between Lake Vänern and the Baltic Sea. When this civil engineering feat was finished in 1832, the total canal system from Gothenburg to Mem on the Baltic became known as the Göta Canal.

The canal leaves the eastern shore of Lake Vänern at Sjötorp and winds toward Lake Vättern, Sweden's second largest body of water. This lake has a north-south length of 130km (80mi) and a maximum width of 30km (20mi). Its remarkable clarity – the lakebed can be seen at depths of up to 10m (33ft) – is due largely to the gravel-purified waters of the natural springs and turbulent mountain streams which feed it. The depth of the lake is considerable: it averages 40m (130ft), but in places reaches as much as 128km (420ft).

A tabletop landscape

An unusual mountain named Kinnekulle rises between Lake Vänern and Lake Vättern. Its summit is a flat tableland, some 14km (9mi) long and 7km (4mi) wide, which looks over Sweden's Lake District from a height of 306m (1,004ft). The landscape of Kinnekulle's slopes differs remarkably from the surrounding patchwork of farms, forests and lakes. The mountainsides rise in a series of flat steps or terraces, separated from one another by sheer cliffs and differing vegetation. On one terrace there is bleak moorland, yet the next is covered by forest.

The change in vegetation is due to the nature of the underlying rock. When, around 500 million years ago, the hard gneiss rock of the region was forced to subside, the land was inundated by sea. Over millions of years, mud, sand and the organic remains of crustaceans, algae and fish, fell to the bottom of the sea and formed layers of sedimentary rock. When the land rose up from the sea bed, around 200 million years ago, volcanic lava welled up through vents in the Earth's crust and settled over local areas of sedimentary rock. Erosion removed softer, exposed layers of rock, leaving behind stepped tablelands, or mesas, of which Kinnekulle is the largest and geologically the most recent.

Each step is composed of a different combination of sedimentary, gneiss and volcanic rock – hence the variation in vegetation which grows on them. The broad diversity of plant types and the presence of rare plants was noted in the mid-18th century by Swedish botanist Carl von Linné (1707–1778), popularly known as Linnaeus and the founder of the modern system of plant and animal classification. He described the region around Kinnekulle and the landscape to the east of Lake Vänern as 'lovelier than any other in Sweden'.

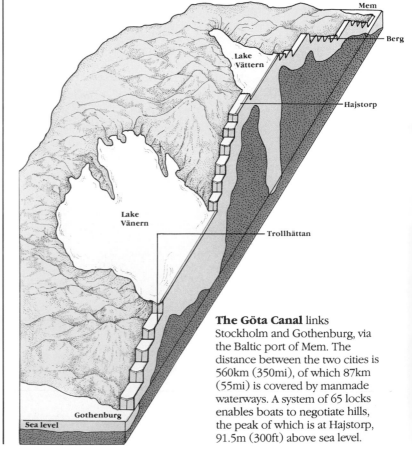

The Göta Canal links Stockholm and Gothenburg, via the Baltic port of Mem. The distance between the two cities is 560km (350mi), of which 87km (55mi) is covered by manmade waterways. A system of 65 locks enables boats to negotiate hills, the peak of which is at Hajstorp, 91.5m (300ft) above sea level.

On Lake Vänern's western shores, the town of Köpmannebro marks the beginning of the Dalsland Canal. This grand waterway runs for 254km (160mi) into Norway and links a number of lakes in Sweden's Dalsland and Värmland regions. The canal was originally built as a waterway to transport freight from ironworks and sawmills in Sweden's Lake District to Norway, and thence to the North Sea.

The lift at Berg, some 40km (25mi) to the east of Lake Vättern, allows boats to traverse one of the steepest gradients on the Göta Canal. It consists of a series of 15 locks – each of which is named after a member of the Swedish royal family – which raises the canal some 36m (118ft) from the waters of Lake Roxen. Eleven locks scale the hillside in single, stepwise flight *(below)* and are regarded as one of the canal's great feats of engineering.

KÖNIGSSEE

Lakeland jewel of the Bavarian Alps

A spectacular ring of precipitous slopes and rugged, snow-capped mountains surrounds with stark grandeur the calm, dark green waters of the Königssee. Pale grey limestone cliffs, devoid of vegetation, hang broodily over the glittering, glacial lake. Dense coniferous forests blanket many of the rocky slopes, and lap the water's edge with an evergreen tide.

The sheer cliffs of the Königssee, which in German means the 'king's lake', rise up like walls and give the lake the appearance of a small fjord. Its ribbon of crystal clear water stretches for about 8km (5mi) and is never more than 1.6km (1mi) across. Lying among the mountains of the Bavarian Alps at an altitude of 602m (1,975ft), the lake has a maximum depth of 192m (630ft).

The best way to experience the alpine beauty of the Königssee is to view it from the deck of a boat. Since 1909, only electrically-powered boats have been permitted on the lake in order to maintain its tranquillity and to help prevent the pollution of its waters. In the 1980s, 21 such boats carry sightseers on round trips from the Königssee's northern shore. At this point, a small stream drains the lake into the River Ache, a turbulent mountain river which rushes through Berchtesgaden and beyond Munich before finally merging with the waters of the River Danube.

Beside the exit of the lake is a quiet bay, the Malerwinkel, which means 'painter's corner'. Here, innumerable artists have captured one of Europe's finest views on canvas. Opposite, the awesome might of the Watzmann group of mountains towers above the Königssee's western flank, to a height of 2,713m (8,901ft). The Watzmann Cliff, unparalleled in the eastern Alps, rises almost perpendicularly to 1,981m (6,500ft) above the lake.

At the foot of this mighty cliff, about halfway along the western shore, a small river has built up a delta. This it has achieved by

The Königssee lies about 5km (3mi) south of Berchtesgaden at the eastern end of the Bavarian Alps in the extreme southeastern corner of West Germany (*left*). About 24km (15mi) south of Salzburg across the border in Austria, the fjord-like Königssee remains one of the few unspoiled lakes in Europe. A wooden house and fishing boat beside an evergreen slope (*right*) are one of many scenes that capture the pristine stillness of the lake.

bringing stones and sand to the lakeside from high mountain valleys. The resulting mound of debris is known as an alluvial fan. On a broad meadow at the edge of the fan stands a small church dedicated to St Bartholomä, or St Bartholomew. Founded in the 17th century by the prince-canons of Berchtesgaden, the church has two 'onion' domes each of which resembles a clover leaf when viewed from above. A 15-minute walk through the Auwald behind the church leads first to the Chapel of St John and St Paul and then to the Eiskapelle, a large, naturally formed vault of ice.

The glacial landscape

A deep, narrow gorge separates the Watzmann Mountains from the Hochkalter which towers to a height of 2,607m (8,553ft). The peak of this mountain is formed by a sharp ridge of rock. On one side of this ridge lies the Blaueis Glacier, the northernmost glacier in the Alps. Its name means 'blue ice' because in bright sunlight its depths seem to fluoresce with a hard, blue light.

At the southern end of the lake stands a massive limestone plateau, the Steinernes Meer, whose name means the 'stony sea'. Measuring 13km (8mi) by 5km (3mi), this enormous shelf is capped by the pyramidal, snowcapped peak of the Schönfeldspitze, which is 2,653m (8,705ft) in height. A pathway from the southern shore of the Königssee, at the foot of the Steinernes Meer, leads to the Obersee. This tiny lake, which is totally surrounded by cliffs and peaks, was formed when an ancient landslide separated its waters from the main lake.

The Blaueis Glacier is a remnant of the ice sheets that carved out the landscape in and around the Königssee. When, more than 25,000 years ago, Europe was in the grip of the ice age, glaciers blocked every valley in the Bavarian Alps. Only the region's highest peaks, such as the Watzmann, rose above the slowly moving rivers of ice. As they moved down the mountainsides the glaciers gouged out massive amounts of rock from the alpine landscape.

When it finally melted around 10,000 years ago, the glacier that carved the Königssee left behind a flat-bottomed, U-shaped valley. After the glacier's power had been exhausted and its ice had receded, a huge pile of stones and gravel was abandoned at the end of the valley. This huge mound, or moraine, blocked off the valley, so trapping the water that formed the lake.

Hidden deep within the underlying rock structures around the Königssee, and untouched by the glaciers of the ice age, lie thick layers of salt. Where this mineral is found near the surface, it has been mined and has formed the basis of a local industry for more than 2,500 years. Even in the middle of the first millenium BC, Celts mined the salt on a large scale and traded not only with neighbouring tribes but also with nations as distant as China.

In the 19th century, this extreme southeastern corner of West Germany had been a hunting reserve for Bavarian monarchs who hunted to extinction all the large predators of the forest – the golden eagle, lammergeier, European lynx, brown bear and wolf. But when, in 1978, the Berchtesgaden National Park was established, it included the region around the Königssee.

The golden eagle (*Aquila chrysaetos*) has since returned to nest in the mountains, and plans are afoot to reintroduce the lynx and the lammergeier. State intervention in the 1930s saved at least two game animals, the red deer and the chamois, from extermination by hunters. But the population of the red deer (*Cervus elaphus*) has expanded 10-fold. Because of its habit of browsing on the leaves and bark of deciduous trees, the red deer is itself threatening some of the parkland's forest, which will demand further conservation measures.

The Bavarian church of St Bartholomä nestles in the shelter of the Watzmann Mountains' sheer cliff face. The Eiskapelle (a natural ice chapel) lies roughly 3.2km (2mi) away through handsome woodland. Electric boats are the only powered craft permitted on this exquisite and secluded lake.

The Königssee does not resemble a lake so much as a secret and self-contained alpine fjord. Its long, narrow contours evoke the shape of the glacier that formed it as one of three valleys lying between the four massifs of Berchtesgaden National Park. Its smaller sibling, the emerald-green Obersee, lies southeast.

The Watzmann Mountains girding the Königssee to the west are in marked contrast to the tranquil waters above which they tower. Rugged and forbidding, their steep slopes bear, among other phenomena characteristic of glacial activity, the bowl-like scars called cirques. At times, only the sharp summits of the Watzmann Mountains can be seen, poking above a soft white collar of cloud.

Although from a distance the massif appears both barren and indestructible, it is home to a rich but vulnerable flora and fauna, now protected by law.

FRASASSI CAVES

Italy's subterranean wonderland

Before the 1970s, few travellers visited the small town of Genga on the eastern slopes of the Appennine Mountains in Italy's province of Ancona. Those who did came to view the 900-year-old castle or to admire the 15th-century triptych in the parish church. But in 1971, the quiet town was catapulted to international fame after a group of men discovered a labyrinth of caves in La Gola di Frasassi, or Frasassi Gorge, some 5km (3mi) to the south.

The gorge, which winds for almost 3km (2mi) between steep walls of rock, was carved by the tumbling waters of the River Sentino, a tributary of the Esino. At its southern end stands the village of San Vittore del Chiuse, renowned for the therapeutic qualities of its sulphurous springs. The limestone walls of the gorge are marked by numerous cave openings. One of these caves, Il Sanctuaria della Grotta, 'the Sanctuary Cave', contains an octagonal church erected by Pope Leo XII in 1828 and an 11th-century chapel dedicated to a local saint, Santa Maria del Frasassi.

The explorers who discovered the Frasassi Caves were members of the Marchigiani speleological group from Ancona, 50km (31mi) northeast on the Adriatic coast. These potholers and cave scientists discovered a cavernous system stretching for 13km (8mi) beneath the Appennine Mountains. A news report of the time called it 'the greatest event in potholing this century'.

In the damp atmosphere of the cave complex, the explorers waded knee-deep through water pools and banks of mud, staring in awe at the surreal stalagmites and stalactites. It was a fantasy world of giant 'marble' pillars, crystalline inflorescences, delicate curtains of frosted rock and huge vaults lined with fragile yet razor-sharp spikes.

The largest cave system in Frasassi's complex is La Grotta Grande del Vento, 'the Great Wind Cave'. Public access to this

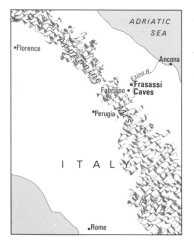

The Frasassi Caves lie within the eastern foothills of Italy's Appennine Mountains, about 20km (12mi) northeast of Fabriano and 50km (31mi) southwest of Ancona (*left*). The intricate cave system opens into the Frasassi Gorge, through which the River Sentino winds before it joins the Esino River, some 3km (2mi) downstream. Many of the rock caverns, such as the Room of the Candles (*right*), are elaborately decorated with stalactites, stalagmites and other sculptures formed from the calcium-based mineral travertine.

subterranean wonderland leads along a smooth walkway that runs for 1.6km (1mi) into the limestone hills. A short tunnel, bored through the rock to make entrance easier, opens into a cathedral-sized chamber. In this cave's centre a shaft, named the Ancona Abyss in honour of the men who discovered the caves, plunges down to unfathomable depths. Close to the hole stands Il Gigante, 'the Giant', an enormous column with ribbed and convoluted sides. Facing the Giant stands the cascading sheet of rock known as La Cascata del Niagara, 'the Niagara Falls', a familiar feature of other limestone caves, such as those beneath Cheddar Gorge in England.

Deeper still lies La Sala delle Candeline, 'the Room of Candles', where an array of short, upright stalagmites emerges from shallow water. White in colour, and encircled at the base by a small 'cup', the majesty of these pillars is further enhanced by imaginative lighting. Illumination also brings out the best from the Grand Canyon formation where as much consideration has been given to shadows as to light. Black areas emphasize cavities and gullies, while bright lights pinpoint the delicate colours of the bands that mark the thin curtains of rock.

The karst landscape

The limestone hills in the Frasassi region of the Appennine Mountains, like those of Guilin in China, are an example of a karst landscape, the term geologists use to describe terrain modelled by acidic rainwater. As rain falls, it absorbs a small amount of carbon dioxide gas from the air. This turns the rain into a dilute solution of carbonic acid which dissolves calcium carbonate, the principal constituent of limestone.

When, around one million years ago, the River Sentino began to carve out the Frasassi Gorge, rainwater trickled and seeped through tiny fractures in the limestone, and gradually widened them. Where the water encountered horizontal fractures, it formed underground streams and excavated long tunnels with connecting caverns. At Frasassi, the acidic water percolated from above and met the underlying water table at the level where the Great Wind Cave now lies. Unable to penetrate deeper, the water spread sideways and fashioned the labyrinthine grotto.

Since their formation the caves have been all but emptied by a drop of almost 300m (1,000ft) in the region's water table. Once this had occurred, stalagmites and stalactites started to form as the water returned to the rocks the calcium carbonate it had earlier removed. Water saturated with this mineral dripped from chamber roofs and cave ceilings to the floors below. Water drops that hung for a few moments before falling were diminished by evaporation, leaving behind minute grains of calcium carbonate, or travertine. Over the centuries, the travertine 'grew' gradually downward to form enormous stalactites.

Dripping water which deposited travertine when it reached the floor formed stalagmites, as in the Room of Candles. In some chambers, such as La Sala dell' Infinito, 'the Room of the Infinite', the stalagmites and stalactites have joined together, linking ceiling and floor with a number of fluted pillars.

Frasassi's flora and fauna, like those found in most other European caves, have become adapted to life in the peculiar underground chambers. The advantages of a high humidity and a constant, year-round temperature, which at Frasassi is 13°C (55°F), is balanced by the disadvantages of total darkness and paucity of food. Yet blind cave salamanders, crayfish, millipedes and flatworms abound in these conditions. Bats are the most prolific inhabitants, roosting by day in La Grotta del Nottole, 'the Cave of the Bats', and emerging at night to feed.

The first group of potholers and speleologists to set eyes on these fabulous dripstone formations in 1971 glimpsed them piecemeal by torchlight. Now the caves are dramatically illuminated to enhance their natural artistry, as well as to allow appreciation of their large size.

The River Sentino emerges from the Frasassi Caves en route to the limestone gorge – La Gola di Frasassi – it has been carving for more than a million years. The gorge has yielded remains of prehistoric life, and is riddled with caves whose Italian names translate intriguingly as Evil Hole, Brown Bear and the like.

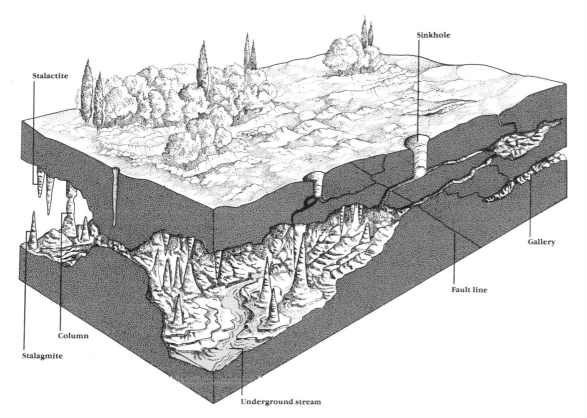

Stalactite

Sinkhole

Gallery

Fault line

Column

Stalagmite

Underground stream

Limestone caves and their fine decorative formations are the handiwork of water and time. Dilute carbonic acid, made when rain absorbs carbon dioxide gas in the atmosphere, dissolves the calcium carbonate of which limestone is composed. Where groundwater finds its way down faults in calcareous rock, sinkholes and shafts can slowly form. As water meets existing horizontal caverns in the rock, these enlarge into galleries and great chambers. Often the openings join up, and elaborate subterranean networks develop.

The stalactites, stalagmites, draperies and flowstone terraces that decorate the caves are known collectively as dripstone. They are formed as minerals in solution, derived from the eroded limestone, are gradually redeposited. This happens when water drips from cave ceilings and walls, and then evaporates.

MOUNT VESUVIUS

The looming peak of destruction

On August 24, AD 79, Pliny the Younger watched the eruption of Vesuvius with horror as he fled with his mother from her house at Misenum on the westernmost point of the Bay of Naples. In a letter to the Roman historian Cornelius Tacitus, the 17-year-old Pliny recalls seeing 'the sea sucked away, and apparently forced back by the earthquake. A dense black cloud was coming up behind us, spreading over the earth like a flood. Darkness fell as if a lamp had been put out in a closed room.'

Throughout the towns of Herculaneum, Pompeii and Stabiae, houses and public buildings were ablaze. Next day the dawn failed to materialize, since the black cloud blotted out the sun. A thick layer of ash coated the ground and the few buildings left standing. Working by the light of torches, people gathered valuables together and fled their burning homes, often shaking the dense ash from their hair and clothing.

The two letters Pliny the Younger wrote to Tacitus provided history with its first eyewitness account of a volcanic eruption. The outpourings of Vesuvius had engulfed three flourishing towns: Pompeii and Stabiae had been blanketed by 6m (20ft) of ash and dust, while Herculaneum was buried by mud to an average depth of 17m (55ft). No one knows exactly how many people died, but later excavations revealed the bodies of more than 2,000 people at Pompeii alone.

A huge, semi-circular crater, known as Monte Somma, cups the cone of Vesuvius, which is formed by the accumulated outpourings of a series of eruptions. The rim of Monte Somma stands 1,280m (4,200ft) above the Bay of Naples and is thought to be the remnant of a prehistoric volcano which blasted itself apart before Vesuvius was born some 200,000 years ago. Between the rim of Monte Somma and the cone of Vesuvius lies a

Mount Vesuvius rises from the western edge of the plain of Campania, about 15km (9mi) east of Naples (*left*). The volcano's western side slopes down to the Bay of Naples where the ruins of Herculaneum were found. The well preserved remains of Pompeii are located around 9km (6mi) to the southeast. The volcano's crater (*right*), which measures about 610m (2,000ft) across and 300m (1,000ft) deep, can be reached via a winding road and by cable car.

deep ravine known as the Valle del Gigante, 'the giant's valley'.

Between AD 79 and 1036, Vesuvius erupted a further nine times before lying dormant for six centuries. On December 16, 1631, another tremendous eruption destroyed 15 villages nestling on the volcano's slopes and killed more than 3,000 people. Lava flowed down to the waters of the bay and Naples was left knee-deep in ash.

The scale of the 1631 disaster prompted the Viceroy of Naples to erect a memorial tablet in the south of the city. On it, he had inscribed: 'Children and children's children. Hear! I warn you now, after this last catastrophe, that you may not be taken unawares. Sooner or later this mountain takes fire. But before this happens there are mutterings and roarings and earthquakes. Smoke and flames and lightning are spewed forth, the air trembles and rumbles and howls. Flee so long as you can.'

Yet such is the stoicism of the people who cultivate the fertile volcanic soil that they ignore all warnings. Seventeen years before the eruption of AD 79, many buildings in Pompeii and Herculaneum had been destroyed by earthquakes. Yet the citizens promptly rebuilt the towns. And despite 19 more eruptions between 1631 and 1944 the local people continued to reconstruct their houses, farms, vineyards and orchards.

The exhumation of Pompeii

During the reconstruction of Resina, a town destroyed in the 1631 eruption, civil engineers digging canals and reservoirs discovered remnants of some Roman buildings in the soil. But they aroused little interest. However, in 1738, when some peasants came across several statues and were able to fetch a high price for them, considerable enthusiasm was stirred. There followed more than a century of looting of the buried towns before Italian archaeologist Giuseppe Fiorelli was commissioned in 1860 to organize a proper exhumation.

The massive layer of ash blanketing Pompeii had preserved much of the detail of town life. As the excavations progressed, Pompeii became more and more like a town stopped dead in time. In addition to luxurious baths and impressive public buildings, Fiorelli's team uncovered 118 taverns, 12 fullers' shops and 10 bakeries. The most astonishing revelation of all was that the bodies of the volcano's victims had left moulds in the solid ash.

Fiorelli dreamed up the idea of pouring plaster of Paris into the moulds and then removing the ash once the plaster had hardened. In doing so he recreated dozens of motionless scenes filled with terrified citizens captured at the moment of their death: a man and his slave loaded down with hundreds of coins and the domestic silver; a sentry trapped at his post; three petty criminals still locked in the stocks.

The 1906 eruption of Vesuvius was the worst since 1631. In an observatory, built in 1845 some 610m (2,000ft) up the side of the mountain, vulcanologists watched the fiery explosions with terror. For 18 days the volcano poured out ash and pumice stone. Villages on the slopes to the north and east were buried, as were 775sq.km (300sq.mi) of cultivated land. The width of the crater increased by about 300m (1,000ft) while the height of its rim fell by 220m (720ft) to 1,277m (4,190ft).

Since the relatively minor eruption of 1944, Vesuvius has lain dormant. The fertile slopes of the mountain are again producing the grapes from which the Lacrima Christi wine is made. Villages and whitewashed farms are dotted across the landscape. More than two million people live in the surrounding country. Yet Vesuvius remains the only active volcano on Europe's mainland and the prospect of another catastrophe is never far away.

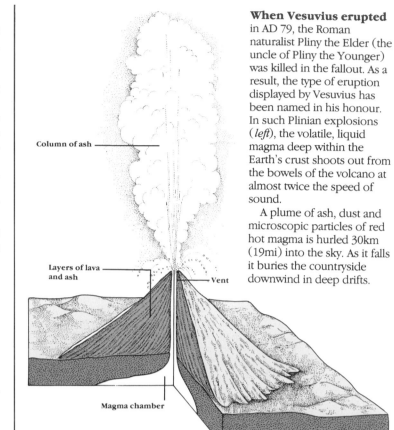

Column of ash

Layers of lava and ash

Vent

Magma chamber

When Vesuvius erupted in AD 79, the Roman naturalist Pliny the Elder (the uncle of Pliny the Younger) was killed in the fallout. As a result, the type of eruption displayed by Vesuvius has been named in his honour. In such Plinian explosions (*left*), the volatile, liquid magma deep within the Earth's crust shoots out from the bowels of the volcano at almost twice the speed of sound.

A plume of ash, dust and microscopic particles of red hot magma is hurled 30km (19mi) into the sky. As it falls it buries the countryside downwind in deep drifts.

An ever-present threat, the volcano of Vesuvius (*above*) looms over the city of Naples. The crater ridge of Monte Somma (the peak to the left of the picture), is all that remains of a prehistoric eruption.

Two thirds of the merchant town of Pompeii has been excavated, including the Forum of Jupiter (*left*). The discovery of rare examples of Roman wall paintings have enabled archaeologists to distinguish four main styles of mural decorations created between 200 BC and AD 80.

Pompeii's tragic victims (*right*) were smothered by a mountainous drift of dust. As the dust hardened it formed a mould around each victim's body before the flesh decayed. By filling the moulds with plaster of Paris, then removing the caked shell, archaeologists have recaptured the moment at which dozens of citizens perished.

43

NAMIB DESERT

Hidden treasures in a fog-bound wilderness

Chill sea winds and dunes wreathed in fog characterize the Namib Desert, located in Africa's southwest corner. The Namib, which fringes the Atlantic coast as a desolate ribbon of contrasting rock, gravel and sand, is truly one of the world's most inhospitable places. In the 1850s, the Swedish explorer and pioneer Charles Andersson confronted its barrenness: '. . . a place fitter to represent the infernal regions could scarcely, in searching the world around, be found. A shudder, amounting almost to fear, came over me when its frightful desolation first broke upon my view. Death . . . would be preferable to banishment in such a country.'

The Namib, which is reputed to be the world's oldest desert, stretches for 2,080km (1,300mi) from the Orange River in the south to the Angolan border in the north. At no point is it more than 160km (100mi) wide, and in some areas it narrows to as little as 10km (6mi). The Kuiseb River, which flows into the Atlantic at Walvis Bay, divides the Namib into two.

To the south of the Kuiseb is a great sand sea containing large areas of parallel dunes with regular troughs, nicknamed 'streets', in between. Here, in the ancient gravel terraces beneath the sands, lies the world's greatest single deposit of gem diamonds. Before the dunes were formed more than a million years ago, the mixture of jewels and gravel was washed down to the sea by the Orange River from South Africa's Kimberley region. The diamonds were swept north on coastal currents, deposited on the shore of the Namib Desert and later covered by sand from the same river valley.

Gravel and rock plains extend north from the Kuiseb River. The treacherous coastal waters of this region regularly snare ships, such as the *Dunedin Star* (1942) and the *Shawnee* (1976), and cast them up on the shifting sands. The rusty hulls of

The Namib Desert fringes the Atlantic shoreline of southwest Africa, from Moçâmedes in Angola through the entire country of Namibia to the Orange River (*left*). The driest of all Africa's deserts, the Namib has two distinct regions: to the north stretch gravel plains and rugged mountains; the south is dunes (*right*), which the wind organizes into a series of parallel lines orientated north-northwest to south-south-east.

many wrecks lie stranded on the beach, which sailors have named the Skeleton Coast, the place 'where ships and men come ashore to die'.

The nature of these westerlies has created the unique environment of the Namib Desert. Flowing north from the waters of the Antarctic is the powerful but cold Benguela Current. Moisture-laden westerlies from the warm Atlantic Ocean are cooled when they meet this current and are forced to release their rain into the sea. As a result, the Namib receives an average annual rainfall of only 2.5cm (1in).

The sand dunes of the desert do receive moisture from another source – from the atmosphere, in the form of fog. Every 10 days or so, a dense night fog rolls inland for a distance of 80km (50mi) or more, shrouding the coast and much of the desert. When it condenses into a thick dew, this fog precipitates an annual average of 4cm (1.6in) of moisture.

Exploiting the desert fog

The Namib, with its unique source of moisture, supports a variety of animals adapted to this unusual environment. Most are small, such as beetles, termites, wasps, spiders and lizards, because only those that can live on a small intake of water survive. In the absence of large predators, the animal species have become largely conspicuous, flightless and defenceless. But strategies for trapping the fog show remarkable evolutionary ingenuity.

The ubiquitous button beetles (*Lepidochora spp.*) excavate tiny parallel furrows in the sand orientated at right angles to the direction of the wind. When the fog rolls in, it condenses on the sand grains of the furrow's ridges: the button beetles then suck them dry. Headstander beetles (*Onymacris spp.*) wait for the fog to arrive on the windward side of coastal sand dunes. When they sense the approach of the white mists, they climb to the crest of a steep-sided dune and balance upside down with their backs to the wind and their heads between their legs. The fog condenses on their carapaces and trickles down into their waiting mouths.

Lizards rely largely on the insects that are their food for their moisture needs. Clown dune crickets and darkling beetles (belonging to the family Tenebrionidae) make succulent feasts. The nocturnal gecko, however, is able to lick the dew from its eyes with its long, flexible tongue. Predator of the reptile community is the sidewinding sand viper (*Bitis peringuei*), a distant relative of North America's sidewinder rattlesnake. Like the sidewinder, this snake skims across the hot dunes, leaving parallel tracks at a 45° angle to the direction in which it moves. When hunting, the snake buries its whole body beneath the sand except for its eyes. Curious lizards who venture too close are injected with poison and then devoured whole.

Among the gravel plains to the north of the Kuiseb River grows the remarkable welwitschia plant (*Welwitschia mirabilis*). Unique to the Namib, and the only species of its genus, the long-lived welwitschia stands up to 2m (6.6ft) tall with two large leathery leaves that are often torn to ribbons by the desert winds. As the desert fog condenses on its leaves, the plant absorbs the moisture via pores on the leaf surface or, when the condensation drips to the ground, through a network of tiny rootlets.

Many of the Namib's invertebrate creatures are detritus feeders, eating debris borne by the hot east winds. These tiny particles of plant detritus or grains of organic matter collect and form a 'larder' in the troughs between the dunes. Beetles, in particular, graze on these food stores either early in the morning or in the afternoon when the desert sands have cooled from their midday high of 66°C (150°F).

Ragged patriarch of the Namib, the welwitschia or tumboa (*Welwitschia mirabilis*) may live for 2,000 years. The sprawling, leathery plant is anchored in the gravelly soil by a woody, carrot-like root, which grows up to 3m (6.6ft) long. This root stores food and water, and provides sustenance in times of drought.

The thirst of a darkling beetle (*Onymacris plana*), one of around 200 species of the Tenebrionid family unique to the Namib, is quenched by fog condensed on grains of sand. The beetle scavenges the dunes for the particles of organic debris that comprise its diet.

The parched clay mosaic at Sossusvlei lies at the foot of the world's highest dunes, many of which reach 300m (1,000ft). It is around 65km (40mi) inland from the Atlantic Ocean, and 300km (180mi) southwest of Windhoek in Namibia, southwest Africa. Sossusvlei is fed a few times each year by small watercourses and underground streams. At these times, animals such as the gemsbok (*Oryx gazella gazella*) come to graze on the temporary abundance of plant life.

The gemsbok has evolved a novel means of coping with the desert heat, which is so intense that it would kill most mammals by destroying their brains. The gemsbok's hot blood cools to normal body temperature before entering the brain by circulating around a fine capillary network in the animal's nose, thus ensuring the creature's survival.

CONGO BASIN

The jungle heart of the Dark Continent

Nineteenth-century travellers referred to the green centre of Africa as the Dark Continent. They thought what appeared to be impenetrable jungle contained ferocious beasts and bloodthirsty cannibals. Its inhospitable reputation, combined with an inaccessibility that defeated all but the most courageous explorers, have in fact made the area relatively safe from man, and thus prevented the destruction of one of the world's greatest rain forests.

The mighty Congo River, with its host of tributaries, flows like a network of arteries and veins throughout the entire primeval region. Portuguese explorers discovered the mouth of the river in the 15th century, but only 208km (130mi) of its lower course were navigable owing to the daunting presence of the Cauldron of Hell, a gorge of cascading, impassable rapids. Anglo-American explorer Henry Stanley (1841-1904) became the first white man to navigate the middle reaches of the Congo after he cut across country from east Africa in 1877–78.

Anglo-Polish writer Joseph Conrad (1857–1924) celebrated the Congo's mystery with his novel *Heart of Darkness* (1902). To him, the Congo was 'A mighty big river resembling an immense snake uncoiled, with its head in the sea, its body at rest curving afar over a vast country, and its tail lost in the depths of the land.' This enormous 'serpent' rises in northwest Zambia as the Chambeshi River, and flows northward and then westward for 4,700km (2,900mi), making it the sixth longest river in the world. The volume of water it discharges is second only to that of the Amazon: from its mouth, 43,300 cubic metres (1,460,000 cubic feet) of water flow into the Atlantic Ocean each second, a figure roughly a quarter of the Amazon's.

The Congo River's basin has a drainage area covering 3,457,000sq.km (1,335,000sq.mi), equivalent to five times the size

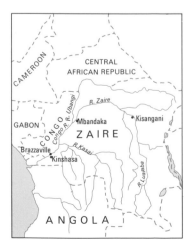

The Congo basin rain forest covers most of the northern half of the country of Zaire (*left*), roughly north of a line from Kinshasa, the capital, to the Rift Valley lakes in the east. The basin of the Congo River, known also as the Zaire, drains almost 13 per cent of the entire continent of Africa. Each tributary of the river (*right*) is an essential branch of Zaire's transportation system: in all, there are 14,166km (8,843mi) of navigable waterways.

of the state of Texas. Much of it lies in Zaire in a vast depression surrounded by mountains and plateaus: from the Rift Valley escarpment in the east to the Cristal Mountains in the west, and from Angola's Lunda Plateau in the south to the Central African Republic's Ubangi Plateau in the north.

To Joseph Conrad, journeying up the Congo River was 'like travelling back to the earliest beginnings of the world when vegetation rioted on the earth and the big trees were kings. An empty stream, a great silence, an impenetrable forest . . .'

The equatorial rain forest of the Congo River basin represents roughly a tenth of the world's total. When viewed from above, this primeval forest resembles an inland sea of rolling green vegetation. In the dense canopy of the forest, some 30m (100ft) above the ground, a rich variety of tree species competes for direct sunlight and heat. Of these, a few manage to outgrow the rest into an emergent layer, opening their entire crowns above the forest. Lianas are quick to take advantage of emergent trees, and woodpeckers make forays for insects amid their branches.

In the canopy habitat grow epiphytes – orchids, ferns and bromeliads that root themselves to the trunks and branches of the trees. Here too live red colobus and moustached monkeys, chimpanzees and mandrills; birds, such as the blue fairy fly-catcher, the African grey parrot and the yellow-casqued hornbill; butterflies, such as the African giant swallowtail and the skipper.

Beneath the canopy other trees stretch upward in search of the scant sunlight filtering through. Here live pythons and vipers; birds, such as the touraco and bee-eater; the epauletted fruit bat and the striped squirrel. In the humid, twilight zone of the forest floor herbivores, including the banded duiker, the bongo and the water chevrotain, graze in natural clearings on leaves and grass. Gorillas feed on buds, berries and stalks, and sleep at night in nests of leaves and twigs. On the ground, termites construct their air-conditioned mounds and, in the thin soil around the boles of tall trees, goliath and wood-boring beetles scavenge for food.

Discovering new animal species

In the first half of the 20th century, three animals unique to the Congo basin rain forest were discovered. The impetus for the discovery of the okapi was based on notes Henry Stanley made about pygmies in 1860: 'The Wambutti knew a donkey and called it "atti". They say that they sometimes catch them in pits. What they can find to eat is a wonder. They eat leaves.' Gradually a picture developed of a large, horse-like animal with the stripes of a zebra, cloven hooves and a habit of browsing at night.

In 1899, the curiosity of Sir Harry Johnston, the British Governor of Uganda, led him into the rain forest where he learned the animal's pygmy name was 'okapi'. When Karl Eriksson, a Swedish officer in the Belgian colonial service, sent Johnston two skulls and a whole skin almost a year later, the governor realized the okapi was a relative of the giraffe. British zoologist Professor Ray Lankester assigned the animal its own genus and gave it the Latin name of *Okapia johnstoni*.

The existence of the Congo peacock (*Afropavo congensis*) came to light in 1936 when American zoologist James Chapin realized that a museum specimen he was studying was not the common peacock by which it was labelled. Instead it belonged to an entirely new genus – the first to be discovered for 40 years. The third new vertebrate species was the fish-eating water civet (*Osbornictis piscivora*), discovered in 1919 in the Ituri forest in the northeast region of the Congo basin. But this chestnut-coloured mammal – reputedly the size of a domestic cat with white facial spots and a black, bushy tail – has never been sighted since.

The African rain forest is made up of six layers, each with its characteristic flora, fauna and microclimate. Descending from the top are: (**1**) the emergent layer, open to the sky; (**2**) the canopy, often of interlocking tree crowns; (**3**) lower tree, the upper part of the complex understory; (**4**) shrub layer, with mature woody plants and young canopy trees; (**5**) the field, of soft-stemmed herbs and seedlings; and (**6**) the floor, a base layer of dead or decomposing vegetation.

A well-established strangler fig (*Commelina diffusa*) will starve and choke its unlucky host. Stranglers start life in the tree tops, where a seed a bird has dropped can germinate in some accumulated soil. The new plant runs roots down along the support tree's trunk to the ground. It then grows quickly, competing with the tree for light and nutrients, until at last it enmeshes the perished host's hollow trunk.

A nine-year-old mountain gorilla (*Gorilla beringei*) – the most endangered of gorilla subspecies – eats wild celery in the protected confines of Volcanoes Park in Rwanda at an altitude of 2,895m (9,500ft). The gorillas feed mainly on the leaves and stems of plants that grow in secondary and montane forests. Loss of habitat is as serious a threat as poachers.

METÉORA

The rocky forest of Greece

On the western edge of the plain of Thessaly in the heart of northern Greece, 24 gigantic rocks rise perpendicularly from the ground. Etched by wind and rain into curious shapes, these ancient pinnacles once echoed to the chants and prayers of the ascetic community of Metéora, home of monks who lived in monasteries and chapels perched precariously upon the rocks.

English traveller Robert Curzon (1810–1873) amply described the gargantuan landscape that had attracted the Byzantine hermits and monks when he visited Metéora toward the middle of the 19th century. '. . . the end of a range of rocky hills seems to have been broken off by some earthquake or washed away by the Deluge, leaving only a series of twenty or thirty tall, thin, needle-like rocks, many hundred feet in height; some like giant tusks, some shaped like pudding-loaves, and some like vast stalagmites.'

Metéora's rocks are composed of a mixture of sandstone and a hard, sedimentary gravel known as conglomerate. These rocks were formed around 60 million years ago as the bed of a sea which covered what is now the plain of Thessaly. A series of seismic movements in the region thrust the sea bed upward to form a high plateau and split the thick layer of sandstone into innumerable faults. Weathering by wind, water and extremes of temperature wore away the fractured sandstone, leaving behind pillars that have been dubbed 'the rocky forest of Greece'.

Greek historian Herodotus, writing in the 5th century BC, records that the local inhabitants believed the plain of Thessaly had once been flooded by the sea, and was fringed by rocky shores. If this were true, it probably referred to an inundation at the end of the ice age, around 8000 BC. Indeed, Metéora's rocks are scoured by horizontal seams which geologists believe were

Metéora stands on the plain of Thessaly beside the village of Kastraki, around 26km (16mi) north of Trikkala and 375km (234mi) northwest of Athens (*left*). The grey monolithic rocks, which stand majestically in front of the Pindhos Mountains, attracted medieval monks seeking isolation. Monasteries, such as Roussanou (*right*), which was built in 1288, are perched almost inaccessibly atop many of Metéora's rocky pinnacles.

created by the waters of a sea. Curiously, neither Herodotus nor any other writer of ancient Greece mentions the rocky pillars of Metéora. This has led to the claim that they did not exist 2,000 years ago, a claim geologists do not take seriously.

Hermits and religious ascetics first inhabited the rocky pinnacles, some of which reach 550m (1,800ft) tall, during the 9th century. The hollows and fissures in the rocks offered the anchorites shelter from the elements, while the sheer cliff walls deterred casual visitors from interrupting their constant soul-searching and prayers. On Sundays and special feast days the hermits gathered together for worship and prayer in a chapel at the foot of the round-topped rock known as Doupiani. By the end of the 12th century, they had organized themselves into a loose-knit community which respected the ideals of solitude.

The Great Meteoron and other monasteries

By the 14th century, the Byzantine Empire, which had ruled the region for more than eight centuries, was beginning to lose its grip. The rich and fertile plain of Thessaly became a battleground as Serbian kings and Turkish raiders vied with each other, and the Byzantine Empire, for supremacy of northern Greece. Peaceful monastic communities seemed particularly vulnerable to the conflicts. In 1334, the monastery of Mount Athos to the southeast of Thessalonica was abandoned. Ten years later, the monk Athanasios led a group of fugitives to Metéora. Between 1356 and 1372, on a pinnacle known as Broad Rock, Athanasios founded the Great Meteoron monastery.

Broad Rock was ideally suited to the reclusive needs of Athanasios and his followers. Once established on the precipitous rock they had complete control of access. A long ladder provided the only route to the top and this could be drawn up by the monks whenever they felt threatened. As the original, simple dwellings became inadequate to house the increasing number of monks seeking refuge in Metéora, larger and more impressive buildings were begun. The ladder was replaced by a net and rope device operated via a windlass from an overhanging gantry.

Visitors to the Great Meteoron, named from the Greek word meaning 'high in the air', were advised to pray as they made the nervewracking ascent. In 1896, a Russian churchman described his trip in the net as 'an agonizing lift, the rope was going here and there all the time dragging me along until at last it reached the top. But as they pulled me towards the wooden platform, they overturned me over the abyss. Horrified, I closed my eyes and nearly lost consciousness.' This precarious method of entry has been replaced by a flight of 115 steps hewn out of the rock.

Of the 24 monasteries that grew up between the 13th and 16th centuries, only five are still inhabited: the Great Meteoron, Agia Triás and Varlaám by monks; Agios Stéphanos and Roussanou by nuns. The influx of tourists in the second half of the 20th century has effectively turned Metéora into a museum piece. The lack of solitude has caused an exodus of older monks and discouraged younger ones from joining one of the monasteries.

The monastery of Varlaám is perched on an obelisk adjacent to the Great Meteoron. The brothers Theophanes and Nectarios built it in 1517, where the 14th-century anchorite Varlaám had his retreat. The two founders established a rigorous discipline which involved praying half the night and mortifying the flesh. The finger of St John and the shoulder blade of St Andrew, both relics reputed to have been housed at Varlaám, bear testimony to this monastery's morbid creed. Frescoes painted in 1548 by hagiographer Franco Catellano adorn the Church of All Saints and depict scenes from the lives of Jesus Christ and the Virgin Mary.

The uninhabited monastery of St Nicholas (Agios Nikolaos) of Anapafsa perches precariously on an outcrop of the enormous Broad Rock on which the Great Meteoron stands. Built around 1388, the monastery was expanded in the first half of the 17th century. In 1527, the walls of its basilica were decorated with richly coloured frescoes by the renowned hagiographer Theophanes the Cretan. Many of the frescoes have been so well preserved that they provide valuable insight into the style and techniques of the artists of the Byzantine Empire.

The pinnacles on which Metéora's monasteries are built stand in the shadow of the Pindhos Mountains. The mighty labyrinth of dark grey stone, which provides an ideal retreat for devout monks in search of peace and solitude, has been fashioned by the sea as well as by wind and weather. Vertical grooves on the rocks were etched by the trickle of rainwater down their sheer faces. Horizontal lines are a memento from prehistoric times, when the waters of a sea that covered the plain of Thessaly lapped incessantly at the rocks.

The simple interior of the monastery of Varlaám (*right*) echoes the humble lives of the monks who inhabit it. Reached by 195 steps hewn into the precipitous rock, much of the monastery shows signs of delapidation. Yet the monks' places of worship – the chapel of All Saints and the chapel of the Three Hierarchs – are adorned with well preserved frescoes and icons of elaborate detail.

PAMUKKALE SPRINGS

Ancient health spa in a fairytale landscape

The startling white cliffs of Pamukkale rise from the fertile plains of Denizli in southwest Turkey like a marble waterfall frozen in time. These cliffs cascade in a widening series of steps from a plateau more than 100m (300ft) high on the slopes of Cal Daği. Vaporous water, loaded with a glistening cargo of minerals, trickles down the cliff face, overflowing from one scallop-shaped pool to the next.

The name Pamukkale, which in Turkish literally means 'cotton castle', has three derivations. First, and most obvious, is the appearance of this natural phenomenon – the dazzling white cliffs do indeed resemble a fairy castle made of cotton. Second, its waters are said to possess the chemical properties needed to clean the cotton produced locally and to make the fabric colour-fast when dyes are used. Third, an anonymous Turkish poet provided the name after he had a vision in which mythological giants called Titans hung out their cotton crop to dry on the side of the mountain.

The cliffs are created by the effects of water which constantly bubbles up from thermal springs under the plateau and emerges into the air at approximately blood temperature, around 38°C (98.5°F). On its underground journey to the surface, this warm water passes through limestone beds and dissolves calcium carbonate from the rock. As it flows over the edge of Pamukkale's escarpment, the water cools and loses its ability to hold the mineral in solution. Consequently, the calcium carbonate, known in its solid form as travertine, gradually and continually precipitates out of solution as it descends the cliff face.

The graded steps on the cliff face are formed by Pamukkale's many water-filled pools. Because the outer margins of a pool cool first, the travertine precipitates at a faster rate here than at

Pamukkale Springs are located 19km (12mi) north of the ancient city of Denizli, and around 250km (156mi) east of Izmir, in the western Anatolian region of Turkey (*left*). Visible on a clear day from across the broad valley of the Menderes River, the gleaming snow-white cliffs cascade down the side of Cal Dagi mountain. As the warm, mineral-rich waters overflow from one pool to the next (*right*), they deposit the white calcium carbonate, or travertine, that has helped earn the cliffs the name of 'cotton castle'.

the centre. As the pools continue to deposit travertine, they grow slowly upward, and after many years create a series of 'columns'. Water that steadily spills over a pool's raised edges trickles down the outside of the 'column', depositing more travertine as a growing outer skin. By constantly adding fresh layers of travertine to the cliffs, the mineral-rich water prevents plants from gaining a foothold and the elements from weathering Pamukkale into a dull, shapeless mass.

Tourists ancient and modern

The abundant hot springs at Pamukkale have been used as a health spa for more than 2,000 years. Even in the late 20th century they are recommended for the treatment of heart diseases, high blood pressure, circulatory problems, rheumatism, eye and skin diseases, as well as nervous disorders in general.

The attraction of Pamukkale as a health spa led to the foundation of the city of Hierapolis, the so-called Holy City. Its ruins, on the plateau from which the waters of Pamukkale spring, have become a modern tourist centre, complete with hotels and swimming pools. Hierapolis was founded in 190 BC by King Eumenes II of Pergamon, a Greek kingdom which controlled much of Asia Minor. The last ruler of Pergamon, Attalus III, knew the Romans were casting hungry eyes over his country. Hoping to avoid the wholesale destruction an invasion would cause, and to keep his family's wealth intact, Attalus bequeathed his kingdom to Rome at his death in 133 BC.

The ancient city of Hierapolis was eventually destroyed by an earthquake in AD 17. Over the next two centuries, the Romans rebuilt it in their own style: with baths, a Temple of Apollo, a colonnaded street and a large amphitheatre that could seat an audience of 15,000 people. Hierapolis became a popular resort for wealthy Romans, and at least three Roman Emperors are known to have visited the city.

Beside the ruins of the Temple of Apollo stands the Plutonium, a paved chamber about 3m (10ft) square dedicated to Pluto, the Roman god of the underworld. A hot stream entering the chamber through a fissure in the rock emanates vapours so noxious that the Plutonium has been dubbed in modern times the Place of Evil Spirits.

According to the Roman geographer Strabo (*c*. 60 BC–AD 21), the fumes were so toxic that they immediately poisoned the sparrows he threw into the fissure. Only the eunuch priests of Cybele, an earth goddess from Phrygia and Lydia in Asia Minor, were said to be immune to the vapours. Such a tradition has led to the conclusion that there must have been an oracle at Hierapolis, since similar, well-documented scenarios exist for the famous oracles of Delphi in Greece and Cumae in Italy.

Behind the Roman theatre, and outside the defensive walls which extend in a rough semicircle around the city, stands the 5th-century Martyrium of St Philip the Apostle. This octagonal church commemorates the death of St Philip who, after retiring to Hierapolis with his daughter, was martyred here in AD 80.

The building that houses the Roman Baths was constructed in the 2nd century AD. It remains remarkably well preserved, largely because the clergy and local inhabitants continually renovated the walls and the roof after it had been converted into a church during the 5th century. Throughout the duration of the Byzantine Empire (AD 330–1453), noblemen and wealthy merchants came to the spa of Hierapolis to bathe and take the waters. But after another earthquake devastated the area in 1334, the city was never again rebuilt. Only in the 20th century has the local population grown and the tourist trade revived.

The ivory-white terraces of Pamukkale cascade down the mountainside like a giant stairway. Each 'step' is a pool of warm, mineral-rich water supported by a wall of fluted, travertine rock. At night, the thermal water floods these terraces; by day, the water evaporates in the heat of the sun, leaving behind further deposits of travertine. The springs of Pamukkale deposit an estimated 4,205 cubic metres (148,500 cubic feet) of travertine each year – enough to cover four soccer pitches to a depth of 30cm (12in).

The scattered ruins of the
Roman health spa of Hierapolis
(*left*) lie beside the modern
Turkish town of Pamukkale. After
an earthquake in AD 17 destroyed
the city founded by the Ionian
Greeks, the Romans rebuilt it in
the grand style with baths, a
theatre and a Temple of Apollo.
The city was abandoned in 1334
after it was ruined by another
earthquake.

Taking the waters

The medicinal value of thermal
spring water comes largely from
its mineral content. The most
important minerals include
calcium carbonate and sulphate,
sodium chloride, iron and
sulphur salts, magnesium
carbonate and magnesia. Gases,
notably carbon dioxide and
nitrogen, often add effervescence
to the potent brew of minerals.

Bathing in warm water is well
known for relaxing the muscles,
but the addition of minerals
brings a curative aspect to the
therapy. Bathing in the
sulphurated waters of spas, such
as Aachen in West Germany, may
alleviate skin conditions. Drinking
the alkaline waters from Vichy, for
example, acts as a purgative. The
carbonated waters of springs,

such as those at Saratoga in USA's
New York State, may relieve
rheumatism and neuralgia. In
general, drinking mineral-rich
water cleanses the alimentary
canal and aids digestion.

RUWENZORI

The elusive Mountains of the Moon

During his stay on the southwest shore of Lake Albert in 1888, Anglo-American explorer Henry Stanley (1841–1904) was blessed with a rare sight. Brooding clouds in the distance suddenly cleared one day to reveal a mountain range previously unrecorded or unseen by white men. 'While looking to the southeast . . . I saw a peculiarly-shaped cloud of a most beautiful silver colour, which assumed the proportions and appearance of a vast mountain covered with snow.'

Stanley had heard tales of the Mountains of the Moon, which the Greek mathematician and geographer Ptolemy (90–168) had claimed were the source of the River Nile. But, like much of darkest Africa in the 19th century, these fabled mountains had not been identified. Rumours linked them with the mysterious and elusive Ruwenzori. As Stanley continued to behold the vision he became 'for the first time conscious that what I gazed upon was not the image or semblance of a vast mountain, but the solid substance of a real one, with its summit covered in snow . . . It now dawned upon me that this must be the Ruwenzori . . .'

In 1906, the Italian Duke of Abruzzi, Luigi Amadeo di Savoia, led an expedition which successfully climbed and mapped the Ruwenzori Mountains. The range, which lies 48km (30mi) north of the equator on the Zaire–Uganda border, contains nine peaks over 4,877m (16,000ft). The highest, Mount Margherita, reaches 5,109m (16,763ft) in height. For an average of 300 days each year, these peaks are all but invisible behind a shroud of thick cloud or dense mist.

The Ruwenzori Mountains form a massif some 120km (75mi) long and 48km (30mi) broad. Unlike all the other high mountains in the east African region, such as Mount Kenya and Mount Kilimanjaro, the Ruwenzori are not volcanic in origin. Their

The Ruwenzori Mountains lie in central east Africa, between Lake Albert and Lake Edward on the border between Zaire and Uganda (*left*). Located on the western branch of the Great Rift Valley, to the north of the equator, the mountain range contains extensive glaciers and glacial lakes. Much of the Ruwenzori's upper slopes are cloaked with mosses, ferns and lichens, together with vegetation (*right*) consisting of giant lobelias, groundsels and heathers.

ancient granite rocks were thrust upward around 2 million years ago when titanic earth movements caused a major subsidence in the adjacent Great Rift Valley.

Wide swamps and marshes based on sediments and debris washed down by frequent rains occupy the valleys and cover the foothills of the Ruwenzori. Here, thick stands of reeds and grasses, including papyrus and *Pennisetum*, grow more than 2m (6.6ft) tall. Elephants (*Loxodonta africana*) push through this vegetation with ease, eating the grasses and startling both natives and travellers with their presence.

Farther up the mountain slopes, in a lush evergreen zone of wild bananas and tree ferns, lives a variety of animals, such as the three-horned chameleon (*Chameleo johnstoni*), an uncommon reptile some 12.5cm (5in) long. Sunbirds, Africa's equivalent of the New World's hummingbirds, drink nectar from lobelias and other flowers, while earthworms as long as 1m (3.3ft) and as thick as a man's thumb weave their way through the moist soil

At about 2,133m (7,000ft), thick stands of bamboo provide cover for leopards, which often follow humans in search of edible refuse. The Ruwenzori's oddest creature is the rock hyrax (*Procavia spp.*), which resembles a rabbit in appearance and shrieks like a guinea-pig. But instead of claws it has the hoofs of an ungulate; its nearest relative is the elephant.

Gigantic plants of the upper slopes
An eerie silence hangs over the higher terrain of the enigmatic Ruwenzori, where rocks and boulders are upholstered by ferns, mosses and lichens. Animals are few and far between at altitudes of around 3,353m (11,000ft), but the plant world seems to run riot. A number of species, which are common and small in temperate climates, here grow to enormous sizes.

Lobelias, such as *Lobelia wollastonii* and *L. bequaertii*, increase their height 20-fold to 6m (20ft). Groundsels (*Senecio spp.*), which are normally 30cm (12in) high, also grow to heights of up to 6m (20ft), their cluster of cabbage-like leaves atop a moss- and lichen-covered trunk. Heathers, usually around 1.2m (4ft) high, grow into trees as tall as 12m (39ft).

The gigantic size of these plants results from the absence of trees in their environment on the mountains' upper slopes. This lack of competition has provided the plants with the opportunity to grow much larger than is normal for their species. Abundant, year-round moisture combined with a mineral-rich, acid soil and high levels of ultraviolet radiation supply the plants with the conditions to achieve such enormous proportions.

The word Ruwenzori, which comes from a local dialect, means 'the rainmaker' – an appropriate name, since the massive, glaciated peaks profoundly affect the weather both locally and throughout the savanna regions of central east Africa. Westerly air currents moving across the steamy rain forests of the Congo basin pick up huge amounts of water vapour. When these moisture-laden winds strike the Ruwenzori they are rapidly forced upward. At these higher altitudes their water vapour condenses to form first water droplets and then ice crystals. This creates the almost permanent cloud cover as well as the local precipitation – around 2m (6.6ft) a year – of rain, snow, sleet and hail.

When strong, steady winds sweep up the mountainsides, the clouds spill over the Ruwenzori and ride above the broad savanna regions to the east where they cause the storms and torrential downpours of the rainy season. Together with meltwater from the glaciers, the abundant rainfall also adds water to tributaries of the Congo River and, via the waterways of eastern Uganda, to the flow of the Nile, the world's longest river.

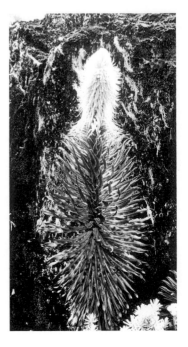

Giant lobelias (*Lobelia telekii*) are among the extraordinary botanical phenomena peculiar to the rain forest belt on the upper slopes of the Ruwenzori, roughly 3,000 to 4,300m (10,000 to 14,000ft) in altitude. Here familiar garden plants grow to tree-sized multiples of their normal height.

A naturalist who, in the early 1960s, chopped down a massive lobelia spike weighing 7kg (15lb) met with superstition from his expedition's porters. None would come near or touch the *mulumbu*, for fear of death.

Mount Speke, named for 19th century English explorer John Hanning Speke, is one of the highest peaks in the Ruwenzori range at 4,890m (16,042ft). It overlooks Lake Bujuku into which drains the meltwater of the Ruwenzori's largest glacier, also called Speke. Here, ice and snow are permanent features, despite the close proximity of the equator. Where the slopes are exposed, the granite mountains owe their fabled silver sparkle to an abundance of mica schists, ancient rocks transformed either by extreme heat or pressure.

The first white man to glimpse the legendary Ruwenzori was Anglo-American explorer Henry Stanley in 1888. A fortuitous parting in the almost permanent cloud cover suddenly revealed the glowing peaks whose existence had hitherto been only rumoured. Even the native Africans were awestruck by the peaks' bright – and so rarely visible – magnificence.

This Victorian engraving (c. 1890) depicts Stanley's vantage point, a village on the shores of Lake Albert, known 100 years later as Lake Mobutu Seso. The African boy who pointed out the mountains to Stanley had no experience of snow, and so said the summits were white with salt.

RIVER NILE

Lifeblood of Egypt's fertility

From Lake Victoria, the Nile's setting changes enormously as this great river flows northward through Africa to the Mediterranean Sea. It winds through forest, desert and swamp; its waters cascade over mighty falls and cataracts. But the Nile is in reality two rivers, the White and the Blue, which merge at Khartoum in Sudan. It is the White Nile, together with the Nile flowing through Egypt, which make up the world's longest river. Its total course of 6,695km (4,160mi) is 255km (160mi) longer than the Amazon's.

'Egypt is the gift of the Nile' wrote the Greek historian Herodotus (c.484–420 BC). Without the Nile's annual flooding the ancient Egyptians could not have grown the food they needed nor established one of the founding empires of western civilization. The farmers who settled along the banks of the lower Nile around 7,000 years ago planted crops in the fertile silt left after the Nile's flood had retreated.

These farmers, who had no idea where the Nile rose, nor why the floods arrived almost unfailingly each year, came to worship the river as a god, whom they named Hapi. A statue in the Vatican Museum in Rome depicts a reclining Hapi clutching ears of corn and surrounded by 16 children, each a cubit in height. The statue symbolized a warning: should the Nile flood fail to reach the height of 16 cubits, around 7.5m (25ft), there would be a famine.

Despite the White Nile's huge catchment area, it is not responsible for the annual flood, since much of its water is lost in the swamps of southern Sudan. As a result, it only contributes about a fifth of the water flowing into Egypt. On the other hand the Blue Nile, which is 1,610km (1,000mi) long, contributes four fifths: every summer, rainfall and melting snows swell this river and cause the annual flooding that, until the Aswan Dam was built in the 20th century, inundated Egypt.

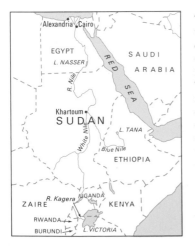

The River Nile and all its tributaries drain almost one tenth of the entire continent of Africa (*left*), a land area totalling around 2,850,000sq.km (1,100,000sq.mi). This drainage basin, which includes parts of nine separate countries, is less than half the size of the Amazon's. Egyptians continue to fish the Nile's waters as they have done for thousands of years (*right*). Boats can navigate the river in Sudan and Egypt throughout the year, except during the low-water season when the cataracts to the south of Lake Nasser become impassable.

A Portuguese Jesuit, Pedro Paez, discovered the source of the Blue Nile at Lake Tana, 1,830m (6,000ft) up in the Ethiopian highlands. A few years later, his compatriot, Father Jeronimo Lobo, described it as being 'two springs . . . each about five feet and a half deep, a stone's cast distant from one another'.

Until the 19th century no one had found the source of the White Nile, and its location remained one of the world's geographical mysteries. The best guess had come from the Greek astronomer Ptolemy who, in AD 150, suggested the Nile drew its waters from the Mountains of the Moon (known today as the Ruwenzori Mountains), situated on the Uganda/Zaire border.

In 1857, the London-based Royal Geographical Society determined to solve the riddle of the Nile's source. Led by explorer Richard Burton (1821–1890), an expedition crossed to Africa from the island of Zanzibar and moved inland, hoping to encounter the Nile and follow it upstream. In the course of their wanderings, Burton and another explorer, John Speke (1827–1864), discovered Lake Tanganyika.

While Burton remained beside this lake, Speke ventured northward and discovered Lake Victoria, claiming it as the source of the Nile. A second expedition, led by Speke, set out from England in 1860 and gathered evidence to substantiate the claim. It was later realized that the only inflow to Lake Victoria, the Kagera River, contributed its flow to the White Nile via a current across the lake's northwest corner. In the 1930s, a German explorer, Burkhart Waldecker, followed the Kagera to its source and hence to the ultimate headstream of the White Nile: 10 springs that trickled into a ravine in Burundi, 96km (60mi) from the shores of Lake Tanganyika.

The course of the Nile

From the northern end of Lake Victoria the waters of the White Nile flow through open scrubland and then tumble 37m (120ft) over the Murchison Falls. The river hurries down to the marshy plain known as the Sudd, an all but impenetrable papyrus and lotus swamp covering around 650,000sq.km (251,000sq.mi).

Below the Sudd, the White Nile becomes broad and stately as it winds its way through barren lands to Khartoum, the capital of Sudan. Here it joins the Blue Nile, so named because its waters are indeed blue; those of the White Nile are, at this juncture, pale green. As it crosses the Nubian Desert to the north of Khartoum, the Nile cascades over four cataracts before flowing into Lake Nasser, known as Lake Nubia in Sudan. This lake was formed by the Aswan High Dam, which was completed at Sadd al Aafi, near Aswan, in 1971. As a result, the Nile's flood waters are used to generate hydroelectricity and to provide irrigation throughout the year, thereby increasing Egypt's food production by more than 50 per cent.

Not all the effects of the dam have been beneficial. The creation of the lake meant that great archaeological treasures had to be transported stone by stone to higher ground. This involved some staggering feats of civil engineering: one section of a mountain containing the temple of Abu Simbel and weighing around 250,000 tonnes was moved 65m (200ft) uphill.

The most detrimental effect has been the loss of fertility to the soil downstream. The huge amounts of silt which the Nile once spread over the fields of Egypt no longer reach them but lie on the floor of Lake Nasser. Farmers must apply fertilizers to keep their land productive. The rich lands of the Nile delta are also starved of the silt on which the farmers depend. As a result, the Mediterranean Sea slowly encroaches, eating away at the fertile land or else drowning it in salt water.

Tissisat Falls in Ethiopia form the most spectacular point in the Blue Nile's course – a furious contrast to the river's calm approach from its highland source at Lake Tana.

The name Tissisat, deriving from local words that mean 'smoke fire', is inspired by the fuming haze in which the falls are shrouded due to the sheer force of their three cataracts. These plunge 43m (140ft) in two stages over rocks of smooth black basalt flanked by riotous vegetation. Their thunder is greatest from September to December, in the wake of the rainy season.

John Hanning Speke, the 19th-century English explorer, made the monumental discovery of Lake Tanganyika with fellow countryman Richard Burton in 1858. Speke carried on alone and discovered Lake Victoria later the same year.

In 1861, Speke returned to Lake Victoria with English explorer and sportsman James Grant and substantiated his claim that the River Nile drew its waters from the lake's water. Speke was able to complete his book *Journal of the Discovery of the Source of the Nile* (1863), the year before his death in a shooting accident.

The Aswan High Dam, inaugurated in 1971, enabled Egypt, for the first time in her history, to control the River Nile's annual flood. It stands 111m (364ft) high and measures 3,830m (12,565ft) along its crest. The dam backed up the waters of the Nile and formed Lake Nasser, which stretches 480km (300mi) to the south.

Huge new areas of land were opened up to agriculture and a hydroelectric power station came on line with a capacity of 2,100 megawatts – enough power to satisfy the energy needs of a city of more than two million inhabitants.

ÜRGÜP CONES

Surreal city hewn from volcanic rock

A peculiar, fairytale landscape greeted French priest Guillaume de Jerphanion when he visited Ürgüp in Turkey's Cappadocia region in 1907. Amongst the mountains and valleys stood a fantastic array of cones, pyramids, needles and honeycombed cliffs. Moreover, these geological sculptures had been excavated, either by the forces of nature or hand of man, and converted into homes or highly decorated churches.

The astonished Father Jerphanion resolved to devote the remaining years of his life to studying the Ürgüp Cones and the rock-hewn churches of Cappadocia. Between 1925 and 1942 he published several volumes of a monumental work which alerted the world to a secret corner of Turkey that had remained virtually unknown since the 13th century.

The congregation of tall, slim, cones rises abruptly from the valley floors to the west of Ürgüp. Most are clustered together, but a few stand in solitary splendour. Each cone consists of a tall pillar of rock, often towering to 30m (100ft) and capped by a black, conical boulder known locally as a 'fairy chimney'. Horizontal bands of red, yellow or white stone define each cone.

Cappadocia's bizarre landscape was formed after two great volcanoes, Hasan Daği to the southwest and Erciyes Daği to the southeast, erupted in a series of outbursts around eight million years ago. The volcanic outpourings covered the surrounding landscape with horizontal layers of lava, ash, cinder and mud. The lava cooled to form a hard, black basalt while the ash coalesced into a soft, white rock known as tufa.

After many thousands of years, the climate became cooler and wetter – more so than it is today. Heavy rainfall produced fast-flowing streams which drained north to the Kizil Irmak river or west to the great salt lake of Tuz Gölü. These streams rapidly cut

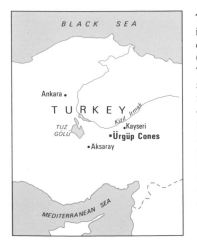

The Ürgüp Cones are located in valleys to the west of the village of Ürgüp, which lies 225km (140mi) southeast of Ankara, Turkey's capital, and 87km (54mi) southwest of Kayseri (*left*). Caves in the soft, volcanic rock of the cones and cliffs (*right*) have been inhabited by peasants and monks for at least 2,000 years. Finely decorated with richly coloured frescoes, more than 300 caves in the region have been transformed during medieval times into simple sanctuaries and churches.

through the soft tufa and created a lattice-work of narrow gorges and steep-sided ridges. Further erosion widened the gorges and generated more intersections. The only parts of the ridges that remained were those protected by weather-resistant basalt boulders. These were whittled away into isolated pinnacles and became the Ürgüp Cones. Their strong bands of colour are due to mineral impurities, such as iron oxides, in the tufa.

Inhabitants of the cones

When viewed from a distance, the outlines of the cones appear smooth and unbroken. At close quarters, however, innumerable doors and windows in the rock faces come into focus. This surreal architecture, fashioned by natural forces, has been adapted by human beings to their needs. Thus caves in cones, and in the cliffs around them, have been inhabited almost continuously for more than 2,000 years.

Peasants transformed the lunar landscape into an agricultural centre while, in the Dark Ages, monks and hermits created an outpost of Christianity. Many caves, still inhabited by Turks, are remarkably comfortable, since the thick rock walls offer protection against the climate's temperature extremes: when snow covers the land, the caves are draught-free and warm; during the long, hot summer days, they become cool refuges from the midday sun.

The arid land around Ürgüp appears barren but with sufficient irrigation it is, in fact, remarkably fertile. Plants thrive in the mineral-rich volcanic soil, enabling farmers to produce abundant crops from their orchards and vegetable gardens. Vines, apricots and peaches, in particular, thrive here. The local white wine has a unique flavour and a bouquet not unlike burning sulphur.

In the 6th century AD, Cappadocia lay under the aegis of the Christian Byzantine Empire (AD 330–1453), centred at Constantinople. At that time, many monks were inspired by the teachings of St Basil the Great (329–379), who had been Bishop of Caesarea (the modern Kayseri, which lies some 87km (54mi) to the northeast of Ürgüp). St Basil established the monastic tradition in Cappadocia when he promoted the idea that monks should live neither in seclusion nor in large communities, but rather in small groups.

At Ürgüp, and in other valleys such as Göreme, monks built small chapels and hermitages in the cones and cliffs. Usually barrel-vaulted with a simple rectangular nave and a small apse, these churches rarely exceeded 8m (26ft) in length. The monks decorated the walls and ceilings with simple designs yet ones symbolic of early Christianity, such as crosses, fish, pomegranates and the palm tree of Paradise. A period of iconoclasm began abruptly in AD 726 when portraits of Jesus and the saints were considered idolatrous and forbidden. But in about 850 realistic scenes and figurative paintings returned to favour, and stories from the Bible were retold on long, segmented paintings.

Around Ürgüp there are more than 150 rock-hewn churches. The most magnificent date from the 10th and 11th centuries when wealthy noblemen vied with each other to establish the best churches or monasteries. They often commissioned the Empire's most skilled artists to decorate the interiors with elaborate, strikingly coloured frescoes.

Politics and war soon brought an end to this craze for Cappadocian churches. In the 11th century, Moslems wrested control of the region from the Byzantine Empire, and although local artists tried to maintain the tradition, they lacked the skill and flair of earlier painters. Following the Turkish invasions of the 13th century, all attempts to build and decorate churches ceased.

The most accessible group of Cappadocian churches lies among the cones of Göreme and in the Göreme Valley (*left*), about 8km (5mi) to the west of Ürgüp. The largest of these is the Tokali Kilise, the Church of the Buckle, which is ornamented with frescoes depicting scenes from the lives of Jesus and the Apostles. Karanlik Kilse, the Dark Church, contains frescoes of the Nativity and the Magi. All around the churches are refuges and hermitages hollowed out of the cliffs and accessible only along narrow passages or via perilous flights of uneven steps.

Honeycombed with human settlements, an enormous rock harbours the village of Üçhisar (*right*), located about 16km (10mi) west of Ürgüp. The view from the summit of Üçhisar, which in Turkish means 'three castles', includes the surreal, volcanic 'city' in its entirety.

Carved out of volcanic rock, the Cappadocian churches are replicas of early Christian architecture. The interior of a church near Ürgüp (*left*) reveals a stark simplicity contrasted with a designer's attention to detail. Columns, arches and domed roofs were included despite the fact that they had no engineering function. In traditional Byzantine style, frescoes depicting scenes from the New Testament lined the walls, ceilings and columns.

DEAD SEA

The saltiest lake on Earth

'Bedouins, pilgrims and travellers visit its shores,' a British visitor remarked of the Dead Sea in the 1930s, 'but these gleams of life only deepen the impression of its unutterable loneliness. The stillness of Death is overall.' The overwhelming quiet is interrupted only by the gentle lapping of the Dead Sea's waters. It seems like a forsaken landscape: no birds cry here and the absence of wind only highlights the lack of trees. The reflections of surrounding hills and mountains disappear from time to time as a ghostly white mist drifts over the water's surface.

Yet the land overlooking the netherworld of the Dead Sea is rich in Biblical history. According to the New Testament Jesus of Nazareth was baptized in the waters of the River Jordan not far from Jericho, the world's oldest known settlement. And from the cave at nearby Qumran the Dead Sea Scrolls were discovered in 1947. In the hills to the east is the fortress of Machaerus where John the Baptist was beheaded; and in the western hills another fortress, Masada, recalls the mass suicide of nearly a thousand Jewish Zealots who refused to surrender to the Romans.

The Dead Sea is the Earth's saltiest body of water. Normal sea water has a salt content of 3.5 per cent. The Dead Sea, with a salinity level of 28 per cent, is eight times as salty. By comparison, the Great Salt Lake in Utah, USA, is six times as salty as sea water. Apart from killing almost every form of life that is swept into the water, the saltiness is responsible for the lake's best known feature: its buoyancy. Sinking and diving are impossible, but it is far easier to swim or float here than in any other stretch of water.

Only a few extraordinary micro-organisms, such as the bacterium *Halobacterium halobium*, can survive in this concentrated brine. These single-celled organisms contain a unique purple pigment, called bacteriorhodopsin, which traps sunlight in a way

The Dead Sea is a landlocked lake lying 24km (15mi) to the east of Jerusalem (*left*) and shared between Israel and Jordan. By the 1980s, it had an area of 1,010sq.km (390sq.mi) – about a third of the size of the state of Rhode Island, USA. It is 75km (47mi) long with a maximum width of 15km (9.5mi). Around 17,000 years ago, its water level was so high it merged with Lake Tiberias to the north. The Dead Sea's waters are so saturated with minerals that salt columns (*right*) appear as strange formations above their surface.

similar to the chlorophyll of green plants. So dependent are they on high salt concentrations that, if the Dead Sea was diluted to only three times the strength of seawater, they would die.

The lowest point on the Earth's land surface, the shores of the Dead Sea are 396m (1,300ft) below the level of the Mediterranean Sea, only 75km (47mi) away. The depression in which these waters sit lies near the northern end of the Great Rift Valley. This rift system is an immensely long trench which zigzags its way 6,500km (4,060mi) from Syria to Mozambique. Its formation began around 25 million years ago when movements in the Earth's crust caused the land to subside.

Mineral-rich waters

In the Dead Sea's northern basin the waters are just under 400m (1,312ft) deep. Those of the southern basin have an average depth of 6m (20ft) although in places they can be as shallow as 2m (6.6ft). Dividing these two basins is a narrow peninsula called Al Lisan, which means 'the tongue' in Arabic.

Around two million years ago, the Mediterranean Sea covered the region. Rock salts were deposited in enormous quantities and, as the Mediterranean receded, these were laid down in the hills and mountains. The Jebel Usdun, a range of hills at the Dead Sea's southern end, are composed of almost solid rock salt. A cap of gypsum and chalk has prevented the infrequent rains from washing away the salt completely. Rising to a height of 150m (500ft), the Jebel Usdun runs for nearly 9km (5.6mi), its steep sides furrowed and chiselled by wind and water.

In Genesis 19, the Bible tells how 'The Lord rained upon Sodom and upon Gomorrah brimstone and fire from out of heaven'. Fleeing the destruction were Abraham's nephew Lot and his family. Lot's wife looked back at the burning cities and was instantly transformed into a pillar of salt. Free-standing pillars of salt are a feature of the Jebel Usdun; and, according to legend, the ruins of Sodom and Gomorrah lie under the waters of the Dead Sea's southern basin.

The River Jordan, together with many small streams, feeds the Dead Sea with water and minerals from the surrounding hills. But the lake has no outlet: its water escapes exclusively by evaporation, leaving behind a concentrated solution of salts, especially magnesium chloride, sodium chloride and potash. Compared with an average annual rainfall of 10cm (4in), almost 2m (6.6ft) of water is lost in evaporation.

Only in the 1920s, when Palestine was under British rule, was the rich mineral content of the Dead Sea regarded seriously as a raw material resource. During World War I, British farms had been starved of the potash fertilizer they had formerly acquired from Germany. Consequently, the British Government sent engineers and chemists to the Dead Sea to discover a method of extracting the minerals on a large scale.

The first chemical plant was built on the northern shores of the lake. But as demand for salt and potash started to outstrip production, a larger complex was built beside the shallow waters of the south. Using evaporating pans as big as a dozen soccer pitches, the minerals were recrystallized one by one from the Dead Sea's water. First to be harvested was salt, then potash and finally bromide. Today, Israel operates large, more sophisticated complexes to extract these minerals, which are essential to the country's glass and fertilizer industries.

The Israelis have plans to dig a canal from the Mediterranean Sea to the top of the hills above the Dead Sea. The benefits would be twofold: to generate hydroelectric power and to top up the diminishing waters of the salt lake.

The Dead Sea's setting, looking south from above, appears as stark as its name. The lake lies in the Negev Desert in the Jordan Valley, an environment virtually devoid of vegetation.

Rain is scarce here and, as in most deserts, is violent when it comes, transforming the dry streambeds into furious torrents. But these unreliable, seasonal deluges have little chance of keeping pace with the lake's high rate of evaporation.

The water is shrinking from the sandy shore. Although the level rose somewhat during the earlier part of the 20th century, climatic changes, as well as the diversion of the Jordan River's headwaters for farming, have caused it to drop again.

A bulldozer shifts great blocks of salt at the Israeli-run Dead Sea Works at Sedom on the sea's southwest shore. The residue left by the evaporation of the world's briniest body of water is a rich source of mineral salts – from common sodium chloride (table salt) to potash, bromine and magnesium – all of which are commercially exploited on a large scale. The chemical harvest, destined for uses as diverse as farm fertilizer and pharmaceuticals, is gathered in oppressive heat; summer temperatures around the lake frequently top 40°C (104°F).

Taking the cure, with knees protectively creamed against the still-strong afternoon sun, a visitor demonstrates the Dead Sea's matchless buoyancy.

Bedouin legends have long extolled the sea's therapeutic qualities. Today the combination of mineral-rich waters, climate, local hot springs and Piloma mud attracts health-seekers in their thousands. Ailments of the skin, joints or respiratory system are commonly treated conditions.

NGORONGORO CRATER

Wildlife sanctuary of the Rift Valley

The precipitous slopes and gullied walls of Ngorongoro's collapsed volcano in the heart of Tanzania enclose one of Africa's most spectacular wildlife reserves. The rich grasslands on the floor of this volcanic crater are a paradise for a permanent population of about 25,000 to 30,000 mammals. Masai warriors, who were evicted from the crater in the middle of the 20th century, revered Ngorongoro; even in the worst droughts they could rely on its springs to quench their thirst.

Ngorongoro is one of many extinct volcanoes in east Africa's Crater Highland region. These volcanoes first erupted around 25 million years ago, at a time when the Great Rift Valley, which runs for 6,500km (4,060mi) from the Zambesi River in the south to Syria in the north, was also being formed. An extensive upheaval in the Earth's crust was caused when two of its immense sections, or tectonic plates, began to move apart in a process known as continental drift. East Africa resides on one of the plates, while the remainder of the continent rests on the other.

As the two tectonic plates drew apart, a rift opened up in the Earth's crust, allowing molten rock, or magma, to escape from the Earth's core. This fiery liquid burst through the cones of Ngorongoro and other volcanoes, showering the surrounding plains with lava and dust. After the eruption, a large lake of molten rock is thought to have formed beneath Ngorongoro but close to its surface.

Around 2.5 million years ago, additional disturbances in the underlying rock structure caused this magma lake to drain, leaving Ngorongoro poised above a huge underground cavity. At this time, the volcano's peak was an estimated 4,570m (15,000ft) above sea level, a similar height to the present-day Mount Kilimanjaro, some 208km (130mi) to the east. Eventually, the weight

Ngorongoro Crater is located in northern Tanzania, some 560km (350mi) northwest of Dar es Salaam, the country's capital, and 240km (150mi) southwest of Nairobi in Kenya (*left*). Short grass covering more than two thirds of the crater floor can support around 250 grazing herbivores in each 2.6sq.km (1sq.mi). Meanwhile, the shallow, salty waters of the crater's lake (*right*) offer rich pickings for a multitude of flamingoes and a playground for lions.

of rock, combined with fresh eruptions, caused the volcano to collapse inward, leaving behind an enormous crater – or, more properly, a caldera.

The floor of the Ngorongoro Crater, which covers an area of about 260sq.km (100sq.mi), lies 610m (2,000ft) below the rim and about 1,830m (6,000ft) above sea level. Round Table Hill, a low, flat-topped hill to the northwest of Ngorongoro's plain, is thought to be the remains of the volcano's peak. The flat basin, which is roughly circular with a diameter of about 16km (10mi), contains grasslands typical of east Africa's savanna.

Unlike the Serengeti Plain to the west, where more than two million animals must migrate when the wet season is over, the Ngorongoro is blessed with an almost continuous supply of water. This natural irrigation maintains the habitat through the year and prevents the herbivorous animals seeking grazing lands beyond the crater's rim. In the dry season, the eastern Serengeti is almost deserted whereas the animal population of Ngorongoro rarely falls below 80 per cent of its maximum. Animals leaving the crater often do so at night via ancient trails forged by their predecessors.

Of flamingoes and herbivores

Two rivers, the Munge and the Lonyokie, supply the crater with water. They feed swamps along their course and eventually flow into the glittering blue waters of a soda lake. This lake lies at the lowest point of the crater but has no outlet. Evaporation tends to leave its waters brackish and, because of this, the aquatic life is restricted – a contrast to other east African lakes, which are rich in wildlife. Lake Tanganyika, for example, contains more species of fish than any other lake in the world.

Huge flocks of flamingoes wade through the shallow waters like a moving pink carpet. At the slightest hint of danger the flocks take to the air and circle majestically before returning to the lake. Two distinct flamingo species feed in the warm waters. The lesser flamingo (*Phoeniconaias minor*), the smaller of the two, feeds on microscopic green algae, which live in the surface waters of the lake. The fine filtering mechanism in the bird's beak strains the water and collects the algae. Lesser flamingoes feed all day long, filtering around 30 litres (6.6 gallons) of water an hour.

The greater flamingo (*Phoenicopterus ruber roseus*) has a coarser filtering mechanism in its beak, which enables it to sift through the mud and sediment on the lake bed for crustaceans, small fish and organic detritus. Not only do the two flamingoes eat different food but they occupy different parts of the lake. The greater flamingo is restricted to the shore where it can reach the mud; the lesser flamingo filter feeds while walking or swimming throughout the whole expanse of the lake.

Munge Swamp to the north of the lake provides a permanent waterhole for hippos and elephants, while during the dry season much of Ngorongoro's wildlife congregates here. As herbivores, such as zebra, wildebeest and gazelles, drink and graze, they are forever on the alert. Their camouflage is only a temporary measure: the black and white stripes of the zebra (*Equus burchelli*) break up its outline; the brindled coat of the wildebeest (*Gorgon taurinus*) matches the browns of the dry savanna.

These deceptions do not fool the most successful of Ngorongoro's predators, the hyena (*Crocuta crocuta*). The flat crater floor, almost devoid of obstacles, such as trees and rocks, provides an ideal environment for packs of hyenas to run and hunt at night. Lions (*Panthera leo*), which hunt by stealth, are less successful in this open landscape. Instead, they help themselves to the hyenas' kill.

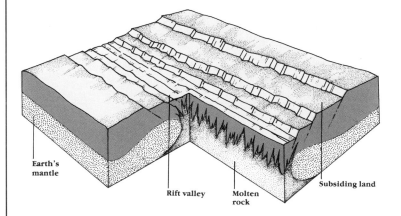

The Great Rift Valley is a huge chain of lakes, ravines, volcanoes and seas in east Africa and the Middle East, stretching from Mozambique to Syria. Much of it has been created by movements of tectonic plates, which fit together in the Earth's crust like pieces in a jigsaw.

In east Africa, two plates pull in opposite directions, tearing the Earth's mantle apart, and causing land along the plates' boundary to subside. In the process, molten rock erupts from deep within the Earth and forms volcanoes, such as the original peak of Ngorongoro.

The Olduvai Gorge lies at the eastern edge of the Serengeti Plains in northern Tanzania. A steep-sided ravine, some 48km (30mi) long and 90m (295ft) deep, Olduvai – like the Ngorongoro Crater to the east – is part of the Great Rift Valley. The gorge was formed by a river that carved its way through various rocky sediments and volcanic deposits laid down over a lake bed in the past 2 million years.

As a result of the pioneering work of paleoanthropologist Louis Leakey (1903–1972) and his wife Mary, a rich harvest of fossils and stone tools have been discovered in Olduvai's rocky layers. These fossils include the bones of 50 early hominids, such as *Australopithecus robustus* and *Homo habilis*, which are some 1.75 million years old.

The open savanna of the Ngorongoro Crater provides ideal grazing ground for herds of herbivores, such as zebra and wildebeest. The crater also supports a large population of black rhino (*Diceros bicornis*). This is unusual, since rhinos prefer to sleep through the heat of the day in the shade of scrub and brush plants, which are hard to find at Ngorongoro. Their success in the crater is probably due to the cornucopia of shrubs, herbs and clover that makes up the greater part of their diet.

MADAGASCAR

Living museum of rare creatures

Since 1500, when Portuguese navigator Diogo Dias became the first European to record a visit to the island, Madagascar has been regarded as a living museum of unusual and unique flora and fauna. The fourth largest island in the world, after Greenland, Borneo and New Guinea, Madagascar measures 1,570km (976mi) long and has a maximum width of 569km (355mi). The island is roughly divided into three north-south zones: a broad central plateau, which rises to an altitude of 1,400m (4,500ft), is flanked by tropical rain forest to the east and mixed, rolling woodland to the west.

Along the uncannily straight east coast, mountains rise sharply to heights of 3,000m (10,000ft). Warm, moist winds sweeping in from the Indian Ocean deposit their rain on the mountain slopes and, in doing so, help maintain a coastal belt of rain forest, some 50km (31mi) wide, which runs almost the entire length of the island. In the west, a band of deciduous forest between 100km (60mi) and 200km (125mi) wide undulates down to the coast. As the broad, winding rivers of the region approach the sea, their brackish estuaries develop into dense mangrove swamps.

The far south of the island presents a stark contrast to this picture of natural fecundity. Sheltered from rain-bearing winds by mountains and cut off from rivers by the lay of the land, this semi-desert region is known as the Land of Thirst. A predominant feature of this arid land is a group of plants known as xerophytes which have a thick, spongy bark and can store large amounts of water to see them through times of drought. Such plants include the baobab tree, which appears to be growing the wrong way up (hence its common name of upside down tree) and the *Pachypodium*, which has a gourd-like bole and stubbly branches about 1m (3.3ft) high.

Madagascar lies in the southwestern corner of the Indian Ocean, around 800km (500mi) from the Mozambique coast in southeast Africa (*left*). The island covers an area of 587,041sq.km (226,658sq.mi), roughly twice the size of the state of Arizona in the USA. The remnants of the island's tropical rain forest grow along the east coast where they fringe sandy beaches, such as those at Maroantsetra (*right*), some 450km (281mi) northeast of Antananarivo, the island's capital.

Forests of a xerophyte known as *Didierea* present a weird and menacing sight. Groups of slim, unbranched stems some 10m (33ft) high are heavily protected with thousands of tiny, razor-sharp spines. These spines prevent most herbivores from browsing on the rows of leaves which grow between the prickles and also trap a layer of air to reduce the loss of precious moisture.

One of the island's unique animals makes its home in the inhospitable *Didierea* forest. This is the sifaka (*Propithecus verreauxi*), which basks in the sun at the tops of the thorny stems; yet it can also leap from tree to tree in the rain forest. The sifaka belongs to the lemur family, whose 22 species are found nowhere else in the world. The largest lemur is the indri (*Indri indri*) which measures almost 1m (3.3ft) in length; its bold, black, white and brown markings provide an effective camouflage in the dappled shadows of the forest. Unlike many other lemurs, the indri has a short, stumpy tail. The most familiar is the ring-tailed lemur (*Lemur catta*), a friendly creature with a long tail ringed in black and white. It, too, lives in the dry, rocky south where there are some trees.

A treasure trove of unique wildlife

Ninety per cent of Madagascar's animals and plants are unique to the island. This extraordinary inheritance is the result of an ancient continental shift. When, around 60 million years ago, Madagascar edged away from the continent of Africa, it became isolated from the rest of the world. The island's animals evolved undisturbed and unthreatened by large carnivores and sophisticated apes. Only the arrival of man less than 2,000 years ago upset Nature's equilibrium.

The catalogue of the island's peculiar animals is enormous. Apart from the lemurs, there are 46 genera of Malagasy birds, the largest and smallest of the world's chameleons, and 148 species of brightly coloured frogs. The insectivorous tenrecs form a mammal family of 30 species which has adapted to a variety of different ecological niches. The greater hedgehog tenrec (*Setifer setosus*) inhabits the dry forest and highland plateaus. When disturbed it rolls itself into a tight ball of sharp spines, exactly like the European hedgehog. The rice tenrec (*Oryzorictes hova*) dwells in the ubiquitous paddy fields of Madagascar where it feeds on invertebrates. Its poor eyesight and its ability to burrow deep underground make it the island's counterpart of the mole.

Human activities have been blamed for the extinction of a number of Madagascar's former inhabitants, including the elephant bird, the pigmy hippo and more than a dozen species of lemur. No one knows exactly how many plant species have disappeared. Estimates in the 1980s put the island's total plant species at more than 10,000; it is also reckoned that there are five times as many tree species in Madagascar as in the whole of temperate North America.

The early settlers who travelled across the Indian Ocean from Indonesia and Malaysia in outrigger canoes brought with them the slash-and-burn technique of agriculture. They chopped down the trees of the forest, set fire to them and planted crops in the ash-covered soil. After a couple of years when the soil was exhausted the farmers moved on to clear a new area of forest.

The rain forests of Madagascar continue to be decimated. By 1985, roughly 150,000 hectares (370,500 acres) were destroyed annually. If this rate is maintained, 99 per cent of Madagascar's forests – about 60,000sq.km (23,000sq.mi) remained untouched in 1985 – will have vanished by the year 2000. In their place will be open savanna and grazing land, or fields under intensive cultivation for rice, spices, vanilla and perfume oils.

The elephant bird

The largest bird the world has known stood nearly 3m (10ft) tall and weighed almost 454kg (1,000lbs). This elephant bird (*Aepyornis maximus*), peculiar to Madagascar but now extinct, was more than 30cm (12in) taller and three times heavier than the ostrich. Flightless and unable to run fast, the elephant bird was a grazing and foraging vegetarian, and with its long neck could crop the leaves off the lower branches of trees.

The elephant bird's early ancestors came from Africa but were stranded on Madagascar when the island started to drift away from the mainland some 60 million years ago. Zoologists believe that *A. maximus* and six other species were hunted to extinction by humans from southeast Asia who colonized the island less than 2,000 years ago.

Giant eggs laid by the birds continue to be discovered: weighing around 9kg (20lbs), or eight times the weight of an ostrich egg, these measure 35cm (14in) in length and have a capacity of 9 litres (2 gallons).

The baobab tree (*Adansonia digitata*) is almost as fat as it is tall, with an average diameter of 10m (33ft) and a height of 12m (39ft). Its bulbous trunk is filled with pulpy matter known as monkey bread and its flowers are pollinated by bats. An inhabitant of arid environments, the baobab grows roots which may spread out for as much as 100m (330ft) in each direction.

The ring-tailed lemur (*Lemur catta*) is an agile tree climber but, unlike other lemurs, also spends a great deal of time on the ground with its distinctive black-and-white tail erect. Troops of between 20 and 40 individuals, most of them females and juveniles, may occupy a single territory. Active by day, ring-tailed lemurs feed largely on fruit, leaves, grass and bark. They also drink tree resin which they obtain by puncturing the bark with their lower incisors.

The aye-aye (*Daubentonia madagascariensis*) is one of the smallest lemurs, with a maximum body length of 44cm (17.5in) and a tail up to 60cm (24in) long. The only species in its family, the aye-aye lives exclusively in the trees of the Malagasy rain forest where it is in danger of extinction. The creature emerges at night to feed on insect larvae, eggs, shoots and fruit. Using the long middle finger of either hand, the aye-aye taps a tree trunk and listens with its bat-like ears for movements of wood-boring insects. On locating one, the aye-aye winkles it out with a finger, or strips the wood with its teeth.

BAND-E AMIR LAKES

Jewels in the foothills of the Hindu Kush

When Mohammed, the founder of Islam, died in AD 632, his father-in-law took over the reins of religious power in Arabia. Ali, the Prophet's son-in-law, who was regarded by many as a more worthy successor, was forced into exile. His absence from the corridors of power led to many tales concerning Ali's whereabouts. One such tale relates how, on a journey to Afghanistan, he fashioned the six glittering blue lakes of Band-e Amir.

Ali arrived in Afghanistan with his faithful servant, Kambar, before the country had been converted to Islam. A local tyrant attempted to capture them in the valley of the Band-e Amir, where the river winds through the western foothills of the Hindu Kush Mountains. A furious Ali escaped by climbing a mountain where, after kicking a rock in the direction of his pursuers, he started a landslide. The fall of rock blocked the river and created a lake. Both the lake and the barrier he forged became known as the Band-e Haibat, 'the Dam of Anger'.

When Ali drew his sword, Zulficar, he severed another rock from the hillside which likewise caused an avalanche and created the Band-e Zulficar, 'the Dam of the Sword'. Kambar, taking a cue from his lord and master, fashioned the third lake by forming the Band-e Kambar, 'the Dam of the Servant'. Slaves freed from the tyrant by Ali created the Band-e Rholaman, 'the Dam of the Slaves'. The Prophet's son-in-law then threw cheese prepared by local women into the river and made Band-e Panir, 'the Dam of the Cheese'. The sixth and final barrier, Band-e Pudina, 'the Dam of the Mint', was formed when Ali threw fresh mint into the river's waters.

Upon his return to Arabia in AD 656, Ali became the fourth Caliph of Islam following the death of his rival, Osman. But five years later he was cut down by the knife of an assassin. His

The Band-e Amir Lakes
shimmer among arid foothills at the western end of the Hindu Kush Mountains, some 80km (50mi) west of Bamian (*left*). The six lakes are clustered near the source of the Band-e Amir River, which flows northward to beyond Mazar-e Sharif and terminates near the Afghanistan-USSR border. The lakes (*right*) nestle in a narrow, sun-scorched valley, their waters trickling over dams made of the limestone-based mineral, travertine.

followers established the Shiite branch of Islam, which continues to flourish in the late 20th century. The Shiites believe that Moslems should be led by a semi-divine man who acts as a mediator between Allah and the faithful. Sunni Moslems, the traditional representatives of Islam, believe the faithful stand face-to-face with Allah without a charismatic leader to intercede on their behalf.

The sparkling blue necklace

The magnificent chain of the Band-e Amir Lakes (Band-e Amir means literally 'Dam of the Prophet') resemble a necklace of lapis lazuli amid the bare, rocky foothills of the Hindu Kush. This mountain range is the second highest in the world (after the Himalayas). It extends 800km (500mi) from northeast Afghanistan to northern Pakistan where its tallest peak, Tirich Mir, is 7,692m (25,236ft) in height. The mountains take their names from the large numbers of Hindus who perished while journeying across the austere mountains to India.

The peaks around the Band-e Amir Lakes rise to around 3,000m (10,000ft) in height and rarely receive any rain. Hemmed in by precipitous cliffs, the lakes sparkle in the mountain sunshine – their deep blue colour is due to the clarity of the air as well as the purity of the water. Each lake is dammed by a long ridge of rock and lies at a different level to its neighbours, thus allowing the river to flow from one to the next.

The Band-e Amir River draws its water from the spring meltwater in the surrounding mountains. This gentle outflow of water seeps into the ground and percolates slowly through the underlying limestone, dissolving its principal mineral, calcium carbonate. Dripping water collects to form underground pools and streams in the caves and passages left behind by the dissolved rock. As they emerge from the hills the streams become the headwaters of the river. Laden with dissolved calcium carbonate and other minerals, the river tumbles into a narrow, twisting valley, no more than 12km (7mi) in length.

The travertine dams

In many other parts of the world, waters bearing calcium carbonate deposit their mineral content as a result of either evaporation, as at Pamukkale in Turkey, or rapid cooling, as at Strokkur in Iceland. At the Band-e Amir Lakes an entirely different process is at work. Streams meander sluggishly between the lakes, allowing intermittent patches of swamp and sand to develop in which willows and scrub trees but also aquatic plants can grow. These plants release a chemical which reacts with the dissolved calcium carbonate and forces it out of solution. Under the intense heat of the sun, the mineral hardens into a pale waxy substance known as travertine.

Over many centuries, the travertine has been deposited around the margins of the lakes, resulting in barriers, or 'dams', which trap water in increasingly large basins. These dams are usually around 10m (33ft) high and some 3m (10ft) thick. As the river fills the highest lake, water trickles over the lake's travertine dam and winds across the sand flats to the next lake, where the process of deposition continues.

Of the six lakes, Band-e Panir is the smallest, with a diameter of approximately 100m (330ft). The largest is Band-e Zulficar, which measures some 6.5km (4mi) in length. When the Band-e Amir River clears the final travertine dam of Band-e Zulficar, it runs freely down the slopes of the Hindu Kush toward the lowlands in the north. But its waters are destined to peter out in the burning wastes of northern Afghanistan near the USSR border.

Tribes of essentially Mongol origin form the sparse but colourful population of the rugged Band-e Amir Lakes area. The Hazara (*top* and *bottom*) include both settled and nomadic groups. In practice, many are semi-nomadic as is necessitated by the terrain – arid steppe land which can support few people or animals, and is useful only as seasonal pasturage.

The Pashtoun (*centre*), with their complex tribal structure, keep to remote mountain regions where they have long eluded both the influence and control of central government.

Shepherds' mud-brick houses, huddled in the shelter of one of Band-e Amir's steep-sided lakes, are found close to the all-important water supply. Donkeys, used for short-haul transport, forage for whatever vegetation they can find at such altitude. Survival in this wild, beautiful but demanding environment is a day-to-day achievement.

A 'frozen' cascade greets riders gathered in winter at the foot of one of Band-e Amir's natural travertine dams. Calcium carbonate dissolved in the water that flows down the stepped sequence of lakes precipitates into spectacular 'ice' formations. But they remain a sight accessible to only a few.

VALE OF KASHMIR

Retreat of emperors: place of peace

Tucked between snowcapped Himalayan peaks and the summits of the Pir Panjal, the Vale of Kashmir sparkles like a precious emerald. One of India's richest and most fertile areas, the vale is graced by colourful fields, tranquil waterways and luxuriant palaces. The 19th-century Irish poet Thomas Moore called it 'the Eden of the Earth'. Its chief city Srinagar, where waterways out-number roads, has been dubbed 'the Venice of the East'.

A 12th-century book from Kashmir, the *Rajatarangini*, tells of the valley's legendary origins. According to this source, the entire valley was once a huge lake where lived the water demon Jalodbhava, whom the gods wanted to destroy. But while he remained in the water, Jalodbhava was invulnerable. The conflict came to an end when Kashyapa, a holy man and a grandson of Brahma the creator god, cut a pass through the mountains at Baramula with a magic sword. The lake drained leaving the demon stranded and defenceless.

The geology of the region lends credence to the legend. When the Himalayas were forced upward by the slow collision of India and Asia, beginning around 60 million years ago, the Earth's crust around them was distorted into many ridges and folds. The fold that became the Vale of Kashmir was 900m (3,000ft) deep, 140km (87mi) long and 32km (20mi) wide. Water from melting glaciers and rain-swelled mountain rivers filled the depression to form a huge lake.

Silt and rocks brought down by the rivers were deposited on the lake bed. Simultaneously, the river draining the lake cut a notch in the surrounding mountains. Because of this deposition and erosion the lake eventually disappeared, leaving a valley covered in sedimentary layers more than 600m (2,000ft) deep. Today, the fertile land is drained by the Jhelum River which

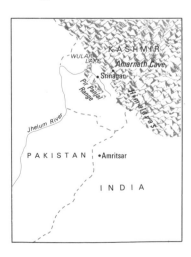

The Vale of Kashmir is located in the northwest Indian state of Jammu and Kashmir, which was partitioned by a ceasefire line after the 1971 conflict with neighbouring Pakistan (*left*). Cradled by mountains at an elevation of roughly 1,650m (5,410ft), the vale is around 135km (85mi) long and 30km (19mi) wide. Symbolic of the vale's renowned fertility, a pear tree blooms in the shadow of Kashmir's surrounding circle of mountains (*right*).

meanders through lakes, marshes and paddy fields, eventually flowing out of Kashmir at Baramula.

At an altitude of over 1,600m (5,000ft), the Vale of Kashmir has a far milder climate than nearby regions in India and Pakistan. It is spared both the bitter cold of mountain winters and the searing heat of lowland summers. And since it is only situated on the edge of the monsoon belt, it does not experience the marked wet and dry seasons associated with the rest of the Indian subcontinent.

The Moghuls' summer resort

In 1526, Zahir ud-Din Mohammed Baber, a descendant of Genghis Khan and Tamerlaine, conquered large areas of northern India. The founder of the Moghul Empire, he was unimpressed by what he found in the new land and wrote: 'Hindustan is a country of few charms. Its people have no good looks, there are no good horses, no good dogs, no grapes, no muskmelons and no cold water.'

When the Emperor Akbar, who reigned from 1556 to 1605, conquered the Vale of Kashmir in 1585 he must have thought it was paradise. Because it was close to the Moghuls' power base in Delhi and near to the trade routes over the mountains, Akbar made Kashmir his summer capital. Its cool climate provided welcome relief from the heat and dust of the Ganges and Indus valleys. Srinagar was transformed into a luxurious summer resort with magnificent palaces and garden sanctuaries, such as Nishat Bagh and Shalimar.

As Jahangir, Akbar's son and father of Shah Jehan who built the Taj Mahal, lay dying his courtiers asked if there was something he desired. 'Only Kashmir', replied the man who was called 'Conqueror of the World'. One hundred years later, the Moghuls were defeated by the Persians and their empire was at an end. But it was not long before another empire, in the form of the British Raj, would have control of Kashmir. However, the British lived not in palaces but in houseboats on Srinagar's waterways because the Maharajah of Kashmir had forbidden Europeans to own land.

In the foothills of the Himalayas

After the British left in 1947, a tug-of-war over who should rule Kashmir started between Hindu India and Moslem Pakistan. The conflict was complicated because the Maharajah of Kashmir was a Hindu while most of his subjects were Moslems. Pakistan invaded from the northwest causing India to come to the Maharajah's aid. A ceasefire was declared on January 1, 1949, but fighting broke out again in 1965 and 1971.

Above the broad farmlands where rice, maize and wheat are harvested, the foothills of the mountains have changed little since the days of the Moghuls. Many areas can only be reached by pony or on foot. The mountains are inhabited by the Gujar and Bakharval tribes, semi-nomads who drive their herds of buffalo or goats to high pastures in the spring and summer.

In the mountains 140km (87mi) to the northeast of Srinagar lies Amarnath Cave. The site of one of the great Hindu pilgrimages, the cave contains a large column of ice formed by water steadily trickling from a natural spring and freezing in the cold air. Hindus believe the ice column is sacred to Shiva, their god of creation and destruction. They also believe the column waxes and wanes with the moon. Around midsummer, especially at dawn on the day when the moon is full, pilgrims arrive from Srinagar in their thousands to throw flowers on the column and to pray to Shiva.

The serpentine procession of Hindu pilgrims winds along a narrow mountain path on its way to Amarnath Cave. Here, an enormous column of ice marks the spot where Shiva, the Hindu god of creation and destruction, revealed the secret of immortality to his wife, Parvati. At their journey's end, pilgrims will pray to Shiva and shower the ice column with gifts of flowers, dried fruit, jewels and silk clothing.

Moghul emperors retreated to the cool climate of the Vale of Kashmir to rest and to enjoy games such as polo, the so-called sport of kings. This mid-17th century painting depicts the game, which has a history of more than 2,500 years, as it was played in Kashmir. The second Moghul emperor, Akbar (1542–1605), who conquered Kashmir, drew up polo's earliest surviving rules.

Saffron, the world's most expensive spice, is harvested by hand from the autumn-flowering crocus (*Crocus sativa*). Kashmir's saffron fields lie at Pampur, some 13km (8mi) southeast of Srinagar on the Jhelum River. Within each purple crocus bloom nestle three orange-red anthers that are pure saffron. More than 4,500 blooms are needed to made 28.3gm (1oz) of the spice, a quantity which is literally worth its own weight in gold.

MOUNT EVEREST AND THE HIMALAYAS
The highest peaks born of ocean floor

Seen from a distance, the pinnacles of the Himalayan Mountains rise like the towers and turrets of a faraway fairytale palace. Snow-capped peaks glitter white in the sunshine as if made of finest marble. Giant pillars of rock seem to flank open gateways. As the sun slips toward the western horizon, its rays bathe the summits in a soft red glow. Shadows chase each other across pink crests. As the light dims and the night thickens, the mountains are fixed as jagged, black peaks outlined against a starry sky.

In the shape of a shallow crescent around 2,415km (1,500mi) long – roughly the distance from London to Moscow – the highest mountain range in the world is between 160km (100mi) and 240km (150mi) wide. Three of the world's largest rivers – the Indus to the north and west, the Brahmaputra to the north and east, and the Ganges to the south – encircle the Himalayas almost completely.

The name of the Himalayas comes from the Sanskrit word meaning 'abode of snow'. The mountains are commonly thought of as a single range, but in fact are composed of three ranges. The lowest and southernmost range, known as the Siwalik Hills, have peaks reaching around 1,500m (5,000ft) above sea level. Farther north are the Lesser Himalayas, around three times as high. Both ranges are patterned with fertile valleys where the climate is mild and many villages thrive. The northernmost range, the Greater Himalayas, contains Mount Everest, the world's highest mountain at 8,848m (29,028ft) above sea level.

Though they tower so high, the Himalayas began their existence on the floor of a sea – fossil fish and the remains of other marine life are often found among the snows. The oceans and continents of the world are carried on immense, constantly moving 'rafts' of rock, or tectonic plates. Around 60 million years

Mount Everest stands at the eastern end of the Greater Himalaya Range, on the border between Nepal and Tibet (*left*). Surrounded by glaciers, the jagged peak of the world's highest mountain punctures the freezing, rarefied atmosphere yet is seldom covered with ice. Mount Everest and its near neighbours, Nuptse and Lhotse, dominate the view even from Gokyo, an alpine region 32km (20mi) to the southwest (*right*).

ago, the plate bearing India moved northward, crushing the floor of an ocean known as the Tethys against the land of Asia. The rocks between them buckled and broke. The ocean floor folded, cracked and was piled up in layer upon layer of disturbed rock. Century by century, uplifted land became mountains and plateaus. These irresistible forces are still at work – geological estimates put the upward progress of the Himalayas at around 5cm (2in) a year.

In 1987, oceanographers analysed sedimentary particles on the Indian Ocean floor which had been washed down from the Himalayas at the time of their conception. They concluded that Mount Everest and the other Himalayan giants had been born around 20 million years ago, making them 10 million years older than had been previously thought.

The lure of the world's highest peak

The first recorded traveller to the Himalayas was Fa-Hsian, a Chinese monk who ventured to the mountains in AD 400 seeking religious truth. Big game hunters from British India, seeking tigers, bears and wild goats, mapped and explored large areas of the mountains. A few hunters, such as B.H. Hodgson in 1832, reported tales of a strange apelike creature, but no specimens were collected. Only in the mid-20th century has this yeti, or abominable snowman, become the focus for scientific investigation. But despite a number of sightings by explorers and mountaineers, and the discovery of huge footprints, the existence of the yeti has not been established.

When Sir George Everest, Surveyor General of India from 1830–1843, led a mapping expedition into the Himalayas he charted many mountains but was unable to pinpoint the highest. In 1852, it was discovered that the mountain known as XV on Everest's maps was higher than its neighbours. In 1865, the mountain was named in honour of Sir George.

Not long after Everest's expedition, the rulers of Tibet and Nepal closed their countries to Europeans. In 1921, the Dalai Lama was persuaded to allow a few Europeans into Tibet. A British party under Colonel Howard-Bury reached the foot of the mountain but had time only to map its lower slopes. In 1924, a junior member of the colonel's party, George Mallory, returned at the head of another team. Watched by colleagues, Mallory and fellow climber Andrew Irvine set out to scale the final peak. The pair had almost reached the summit when they were enveloped by cloud and never seen again. No one knows whether or not they conquered Everest, but new evidence unveiled in the 1980s has led many people to believe they did.

By 1953, the Tibetan side was again closed to climbers but the borders of Nepal were open. In that year, a British expedition, organized with military efficiency by John Hunt and carrying oxygen and tackle developed in World War II, faced the terrible grandeur of Everest.

On the morning of May 29, 1953, New Zealander Edmund Hillary and Sherpa Tenzing Norgay, prepared for the final assault. Setting out from their advance camp at 8,450m (27,900ft), they faced a difficult climb along a narrow ridge: on either side lay a drop of 3,300m (10,000ft). Five hours later, Hillary realized they had reached the summit. 'My initial feelings were of relief,' he later wrote, 'no more ridges to traverse and no more humps to tantalise us with hopes of success. I looked at Tenzing . . . and there was no disguising his infectious grin of delight.' Since this triumphant conquest, the desire by mountaineers to touch the roof of the world has intensified – more than 130 ascents have been successful, five of them without the aid of oxygen.

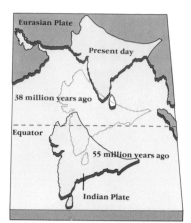

The continents of India, Australia and Antarctica once formed a supercontinent known as Gondwanaland around southern Africa. Some 150 million years ago, the tectonic plate (a giant 'raft' of the Earth's crust) bearing India broke away. It then headed northward across the sea known as the Tethys. When the Indian Plate collided with the Eurasian Plate, some 60 million years ago, the sea bed buckled upward. Thus began the geological process that would create the Himalayas.

Edmund Hillary (*left*), a New Zealand beekeeper, and Tenzing Norgay (*right*), a Sherpa guide on his seventh trip to the mountain, pose triumphantly after their conquest of Everest on May 29, 1953. Sherpas, whose strength on Himalayan expeditions has earned them the name of 'tigers of the snow', have long revered the peak as Sagarmatha, the 'Goddess of the Universe'.

Abominable snowman country surrounds the Tso Rolpa glacial lake in the Upper Rowaling Valley, roughly midway between Mount Everest and Katmandu, Nepal's capital. This characteristic Himalayan territory was explored for signs of the legendary yeti in 1960.

The smoky-grey snow leopard (*Panthera uncia*) haunts the rocky crags and desolate ridges of the Himalayas and other central Asian mountain ranges. A highly endangered species, the solitary snow leopard spends the summer above the snow line, sometimes as high as 5,500m (18,045ft). In winter, it descends to about 2,000m (6,600ft) while tracking its migrant prey, such as the ibex and wild sheep as they forage in forest and scrub.

LAKE BAIKAL

The deepest lake in the world

Until 1891, when Tsar Alexander III of Russia ordered the construction of the Trans-Siberian railway, the remote region of Lake Baikal was familiar only to Siberia's Tunguz and Eventi tribes. The arrival of the railway encouraged fishing and lumber industries to grow. With it also came the realization that Baikal is one of the most spectacular lakes in the world.

Shaped like the first crescent of a new moon, but with its outline broken by numerous bays and peninsulas, Lake Baikal is 636km (395mi) long with an average width of 48km (30mi). The surface area of the lake is around 31,500sq.km (12,200sq.mi) or roughly the combined size of the two US states of Massachusetts and Connecticut. Yet while it is only two fifths of the size of North America's Lake Superior, Lake Baikal is four times as deep. In fact, its maximum depth of 1,620m (5,314ft) makes it the deepest lake in the world.

Gigantic fault lines in the Earth's crust scar the centre of the Asian continent. When, some 80 million years ago, tremendous upheavals and earthquakes widened the faults, a portion of the Earth's crust collapsed and formed a deep, sheer-sided chasm, known as a graben. At the same time, mountain ranges were pushed up on either side. The graben remained predominantly dry, since the water draining into it soon evaporated. But around 25 million years ago, the climate turned wetter; rainfall outstripped evaporation, so starting the long process of filling the lake.

In modern times, more than 300 rivers pour their waters into Lake Baikal while only one, the Angara, drains it. The lake's total quantity of water amounts to 23,000cub.km (5,500cub.mi), or one fifth of the Earth's entire volume of freshwater. And the lake is growing larger. In 1862, a massive earthquake struck the area around the mouth of the Selenga River, which provides the lake

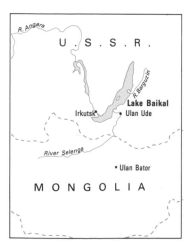

Lake Baikal is located in southeast Siberia in the Buryat Autonomous Soviet Socialist Republic, some 80km (50mi) from USSR's border with Mongolia (*left*). The lake's drainage area of 540,000sq.km (208,494sq.mi) is 13 per cent greater than the combined drainage area of North America's Great Lakes. When the *sarma* wind blows at its usual 130kph (81mph) from the northwest, the waters of Lake Baikal (*right*) form waves more than 5m (16.5ft) high.

with 50 per cent of its water, and severed roughly 175sq.km (77sq.mi) of land from the shore. Water rushed in from the lake and created a new bay, known as Proval Bay. In the long term, the whole of Asia is being reshaped. As the land continues to slip beneath Lake Baikal, it is opening up a still mightier chasm which, in a few million years, will link up with the Arctic Ocean and divide Asia into two.

Baikal is not only the world's deepest lake, it is also the world's oldest. For more than half of its 25 million years, the lake has provided almost constant living conditions and so contains an enormous assortment of endemic life forms. Around 35 per cent of the lake's 600 plant species and 65 per cent of its 1,500 animal species are unique to Baikal's water.

Animals of the deep

All Baikal's creatures depend on the food and oxygen which the algae and plankton produce in the top 50m (165ft) of water. Animals living in the depths of the lake either feed, like shrimps, on the detritus falling from the surface layer, or else eat each other. All but one of the lake's 255 species of freshwater shrimps – a third of the world's total – are found in its deep waters. Moreover, these species belong to 35 different genera, only one of which is found elsewhere in the world.

More than 50 species of fish, half of them unique to the lake, inhabit Baikal's waters. The largest is the sturgeon (*Acipenser sturio*) which grows 1.8m (6ft) long and weighs more than 100kg (220lbs). Because of its prodigious caviar and sought-after flesh, the sturgeon was almost eradicated from Baikal's waters; only strict fishing laws have enabled it to flourish once more.

Perhaps the most unusual fish is the golomyanka, which is represented by two endemic species – *Comephorus baicalensis* and *C. dybowski*. Completely transparent, these scaleless fish grow to about 20cm (8in) in length and accumulate about a third of their body weight as oil. At night, they migrate from as far down as 500m (1,650ft) to feed in the surface waters on the minute creatures that comprise the zooplankton. They must descend again before the temperature of the lake rises above 7°C (45°F), otherwise their oils begin to liquefy and they die. Female golomyanka do not lay eggs but give birth to as many as 3,000 live larvae – an event they rarely survive.

A local Baikal delicacy is the white, salmon-like omul (*Coregonus autumnalis migratorius*). Forming up to 70 per cent of the commercial fish catch of the lake, the omul grows 30cm (12in) long and weighs around 454g (16oz). These fish remain inactive until the summer when the temperature of the lake's surface water rises to about 16°C (60°F). At dusk, and again at dawn, schools of omuls feed voraciously on the zooplankton – which swim to and from their regular night-time feed at the surface.

Every year the lake freezes over by the end of January and remains frozen for four or five months. During exceptionally cold winters the ice can be as thick as 1.2m (4ft). The Baikal or nerpa seal (*Phoca sibirica*), which is unique to the lake, survives the winter by making air holes in the ice which it keeps open by gnawing them from below. In the summer, the seal lives among the rocky crags at the northeastern end of the lake where it thrives on Baikal's abundant fish population.

The Baikal seal, like the omul salmon, is an outsider. Its nearest relative, the Arctic seal, lives 3,200km (2,000mi) away in the Arctic Ocean. But since there is no evidence of an ancient sea in the region, the Baikal seal must have travelled to the lake against the currents of the Yenisey/Angara river system, a voyage thought to have taken place some 12,000 years ago.

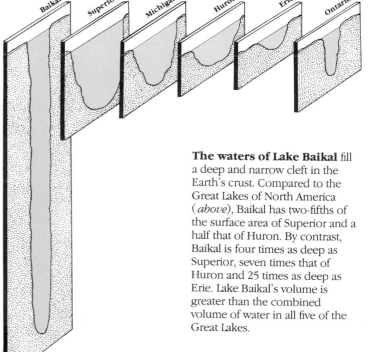

The waters of Lake Baikal fill a deep and narrow cleft in the Earth's crust. Compared to the Great Lakes of North America (*above*), Baikal has two-fifths of the surface area of Superior and a half that of Huron. By contrast, Baikal is four times as deep as Superior, seven times that of Huron and 25 times as deep as Erie. Lake Baikal's volume is greater than the combined volume of water in all five of the Great Lakes.

The rugged terrain
surrounding Lake Baikal contains
a diversity of habitats, including
taiga, tundra, peaty meadows and
various types of forest. In these
environments live 220 bird
species and more than 40 species
of mammal, such as the musk
deer, reindeer and sable. Much of
the land has been preserved in
the Baikalsky and Darguzinsky
State Reserves, which cover a total
area of 4,640sq.km (1,792sq.mi).

The Baikal seal (*Phoca
sibirica*), which is unique to the
lake, is one of the world's two
freshwater species of seal.
Estimates in the 1980s put the
population of Baikal seals at
around 70,000 and on the
increase. They breed on the
frozen lake in winter, giving birth
to pups in solitary snow lairs from
late February to early April.

KRAKATOA

The volcanic eruption that shook the Earth

When the island of Krakatoa erupted on August 27, 1883, it destroyed 300 villages and killed 36,000 people. Houses were cracked open 160km (100mi) away. The sound was heard thousands of miles distant. The shock wave in the air blasted around the globe seven times. Bodies and wreckage were seen floating in the sea for days afterward. It has been called the greatest natural explosion in history.

At the beginning of that year, Krakatoa seemed an ordinary volcanic island, lying in the Sunda Strait between Java and Sumatra in what was then the Dutch East Indies and is today Indonesia. Measuring 28sq.km (11sq.mi) in area, the island was dominated by a central peak 820m (2,700ft) high. Few islanders worried about the volcano – there had been no sign of activity since the mountain had erupted two centuries earlier, in 1681. Some even thought the volcano extinct.

On May 20, 1883, the mountain's cone burst into life, hurling hot ash into the sky but dying down soon afterward. As the summer advanced, several other small eruptions followed. Still there were few who worried – such minor eruptions happened all the time in the islands. As August passed, loud groans were heard deep underground like a giant beast awakening.

Early in the evening of August 26, a deafening explosion rocked the island. The central cone erupted violently, throwing a column of dense ash and smoke 27km (17mi) into the air. By the early hours of the next morning, many of the islanders had taken to sea. A lone Englishman who escaped later remembered the crowds unable to leave: 'The poor natives, thinking that the end of the world had come, flocked together like sheep, and made the scene more dismal with their cries.'

Sixteen kilometres (10mi) distant, Captain W.J. Watson of the

Krakatoa Island lay in the Sunda Strait roughly midway between Java and Sumatra (*left*) before it destroyed itself in a massive volcanic explosion in 1883. All that remained of the shattered island were tiny jagged islets covered in volcanic debris. In 1952, a new island rose from the graveyard of the old. Known as Anak Krakatau, or 'Son of Krakatoa', this young active volcano stands more than 150m (500ft) above the water (*right*).

Charles Bal, grew seriously worried – as bolts of lightning struck the island, St Elmo's fire lit up his ship. He later wrote that 'the intense blackness above and around us, broken only by the continued explosive roars of Krakatoa, made our situation a truly awful one, some of the pieces falling on us being like iron cinders.' He had also noticed how warm the sea was to the touch.

The fatal blast

At 10 a.m. on August 27, a cataclysmic explosion ripped the entire island apart. Two-thirds of Krakatoa simply ceased to exist. More than 19cub.km (4cub.mi) of rock were pulverized to dust and thrown into the air. This is around four times the volume expelled by the explosion of Mount St. Helens, USA, in 1980. Dust and stones were catapulted 55km (34mi) upward into the stratosphere. Within minutes the sky went dark around the island. Before long, an area 280km (180mi) across had been plunged into total darkness.

The sound of the blast was stupendous. In Batavia, nearly 160km (100mi) away on the north coast of Java, people in the street were temporarily deafened. The islanders of Celebes (the modern Sulawesi), nearly 1,600km (1,000mi) to the east, thought they heard distress rockets and launched their sea rescue services. On the island of Rodriguez, more than 4,800km (3,000mi) to the west in the Indian Ocean, the people imagined a naval battle was being fought just beyond the horizon.

The gaping crater left by the blast was 6.4km (4mi) across, but had been plunged 275m (900ft) beneath the sea. The surrounding waters rushed in with such force they created a massive tidal wave, or tsunami, more than 30m (100ft) high. The wave sped outward from the island at 1,120km (700mi) per hour, almost the speed of sound. It crashed ashore on neighbouring islands and coasts sweeping everything before it.

Beside the town of Anjer, 24km (15mi) away on the coast of Java, a Dutchman named de Vries saw the wave approaching like 'an enormous mass of water, mountain high'. Running for high ground, de Vries climbed a coconut tree and waited. The wave struck the town and buried it. He recalled: 'I looked round. A fearful sight met my eyes. Where Anjer had stood I saw nothing but a foaming rushing flood . . .' Other shores of Java and Sumatra met a similar fate. With reduced force the wave rolled ashore in Australia and California.

On August 28, Captain T.H. Lindeman sailed his steamer *Gouverneur-Generaal Loudon* along the north Java coast to Batavia. He described the aftermath of the wave in his log: 'Everywhere the same gray and gloomy colour prevailed. The villages and trees had disappeared; we could not even see any ruins, for the waves had demolished and swallowed up the inhabitants, their homes and their plantations. This was truly a scene of the Last Judgement.'

The dust remained in the atmosphere for many months afterward, creating beautiful sunsets and blue moons. All around the globe fantastic shades of red, purple and pink illuminated the night sky. It took three years for the dust to settle completely.

Most of the pre-eruption island of Krakatoa vanished, leaving several islands and islets in the sparkling blue seas. In 1927, renewed volcanic activity built up another island beneath the sea's surface. An explosion in 1952 thrust it into the open air. Named Anak Krakatau, meaning 'Son of Krakatoa', the tiny active volcanic island is 150m (500ft) high. Lying at the centre of four islands, whose jagged outlines rise above smooth blue waters like broken teeth, the young Anak Krakatau seems to be a sign that another volcanic cycle has begun.

A vivid engraving based on a photograph taken shortly after Krakatoa's eruption offers a glimpse (*above*) of the island before much of it sank beneath the waves. The billowing column of dust and stones reached a height of some 55km (34mi).

The tidal wave set in motion by Krakatoa's explosion struck many Indonesian coasts. A late 19th-century engraving (*below*) shows the fate of a steamer beached by the wave at Telekbetung, some 24km (15mi) north of Krakatoa.

The island of Anak Krakatau
(*above*) rose from the sea some
69 years after Krakatoa had almost
obliterated itself. Both the new
island and the remnant of the old
were quickly colonized by plants
and animals. Botanists who visited
Krakatoa in 1886 discovered more
than 30 species of plants; ten
years later, the island was covered
in grasses and shrubs.

Grass grown from seed borne
on the wind or brought in by
birds bring life to the dark soil of
Anak Krakatau. Yet hardy grasses
are not usually the first colonizers
of new volcanic islands – they are
preceded by lower plants, such as
blue-green algae, lichens, ferns
and mosses.

GUILIN HILLS

Sugarloaf tors etched with caves

Chinese artists and poets have, for many centuries, celebrated the spectacular landscape around the city of Guilin. Here, rounded, steep-sided pinnacles of limestone, which rise abruptly from the flat plains either side of the Xi River, are often shrouded in mist or wrapped in a thin tissue of white cloud. The T'ang dynasty poet Han Yu (768–824) described the river as 'a turquoise gauze belt' and the hills as 'a deep jade hairpin'.

The turrets of Guilin are spread out along 48km (30mi) of the Xi River and soar to heights of 100m (330ft) or more. They are studded with craggy outcrops, vertical fissures and parallel ledges of rock, where stunted trees with dark green foliage precariously survive. Long, sturdy vines trail from the branches of the trees, either dangling their tendrils over cliffs or else creeping over rocks. Orchids and other wildflowers add bright splashes of colour to the otherwise grey-green scenery.

Around 300 million years ago, the region was covered by an ocean which had a bed of a tough, resistant rock known as quartzite. Over succeeding millenia, layers of sediment were deposited on top of the quartzite. At first, the deposits were of fine mud which, in time, become converted into a deep layer of shale. Later sediments contained large concentrations of calcium carbonate which were gradually transformed into limestone. Upheavals in the Earth's crust caused the ocean floor to rise above the level of the sea, at once exposing the thick layers of limestone to the erosive power of the elements. Vast quantities of soft, porous limestone were dissolved and washed away, leaving behind the more resistant pinnacles to tower above the plain.

The Guilin landscape is renowned among geologists as one of the world's prime examples of tower karst, a landform based on limestone that is riddled with caves, underground streams and

Guilin is located in the north of China's Guangxi-Zhuang Autonomous Region, some 545km (340mi) northwest of Hong Kong and 480km (300mi) southwest of Changsha, the capital of Hunan Province (*left*). Recognized as one of China's premier beauty spots, the mist-shrouded, sugarloaf tors around the city of Guilin rise from the plain like dragons' teeth (*right*). An inspiration to poets and artists for more than 1,300 years, Guilin's bewitching hills have become the epitome of a Chinese landscape.

passageways. The sole architect of the entire region of Guilin is rainwater, made slightly acidic by the addition of carbon dioxide from the air. As rain falls, minute amounts of the gas become dissolved in it, so converting it into a weak solution of carbonic acid.

The acidic water works its way into any crack, fault or fissure in the rock and etches away the limestone. Little by little the openings are widened into passages or caves and initial trickles of water become streams. This erosive agent is abundant around Guilin where the average annual rainfall is between 1.1m (3.6ft) and 2.8m (9.2ft). Combined with the subtropical humidity and warmth – the average July temperature is 26°C (79°F) – the chemical power of the water is maximized.

The lyrically-named caves and hills

Guilin's water is creative as well as erosive. Rainwater that falls on the region becomes the vehicle for calcium salts dissolved from the rocks. This limestone-rich water steadily seeps through the roofs of many caves and deposits its load of calcium carbonate, either as stalactites or stalagmites. Hundreds of stalactites hang down from the ceiling of the Reed Flute Cave, lying 6km (3.7mi) north of Guilin, where they form a miniature counterpart of the hilly landscape outside.

In the 1930s, and again during World War II, the caves of Guilin were used as air-raid shelters by hundreds of thousands of refugees from Chinese provinces to the north. Indeed, Reed Flute Cave had long acted as a secret hideout specifically for the local inhabitants and was only opened to the general public in 1958. Near the entrance to this cave stands the Old Scholar, a stalagmite resembling a seated scribe. According to an old tale, this is a poet who sat down to write about the beauties of the cave but turned to stone before he could find the right words.

The aptly-named Breezy Cave runs completely through the Piled Silk Hill, which lies beside the Xi River to the north of Guilin. The peculiar structure of the cave and the positions of the surrounding hills, are instrumental in producing the cool breeze that continually plays through the tunnel, no matter how still or sultry the weather outside.

The name Guilin means, literally, 'forest of cassia' and refers to the abundant Chinese cinnamon trees (*Cassia lignea*) that grow in the region. Between August and October, the redolent blossoms fill the city with a rich and gentle fragrance. The city centre is dominated by the Peak of Solitary Beauty, which lies within the ruins of a Ming dynasty palace built in 1372.

The Peak of Solitary Beauty offers a panoramic view of the eroded limestone hills, which may be cloaked in mist or reflected from the calm waters of the river, lakes or paddy fields. The natural features of many tors have, in traditional Chinese style, inspired their lyrical names: Bat Hill, Camel Hill, Five Tigers Catching a Goat, Climbing Tortoise, Elephant Trunk Hill.

In 1973, the Chinese authorities opened up the city of Guilin and the scenic stretches of the Xi River to foreigners. Factories polluting the river were replaced by landscaped gardens and modern hotels. Ninety thousand cassia trees were planted. Within five years, more than 50,000 foreigners visited the city, almost two thirds of them Chinese expatriates.

Many visitors cruise along the Xi River in state-run boats which must navigate the numerous shoals and rapids. A local tale maintains that boatmen drowned in these waters are turned into devils who try to overturn other boats and drag the occupants below. An intriguing sight on the Xi River are cormorants trained by fishermen to perch on the stern of small boats and dive into the water whenever they see a fish.

The quaint but still effective practice of cormorant fishing contributes to the storybook atmosphere along the River Xi. Against the ethereal backdrop of Guilin's rounded pinnacles, fishermen exploit the birds' specialized hunting and diving skills, fitting tight collars around their necks to prevent them from swallowing any fish they catch other than small fry.

The technique does not always pay dividends in the spring months. Then, when snow meltwater clogs the river with large quantities of mud and sediment brought down from the mountains, even the sharp-eyed cormorants are hard pressed to spot their prey in the murky river water.

The great or black cormorant (*Phalacrocorax carbo*) is both the largest and the most widely distributed of some 30 species in its family. The birds dive for their prey – mainly fish, but also crustaceans and amphibians – staying under water for about half a minute. They wait until they have resurfaced to eat, and give their catch a good shake before consuming it.

Spectacular caves add an extra dimension of mystery to the unique Guilin landscape – one of the world's premier regions of tower karst. Yawning from the base of many of the strange conical hills are the entrances to caverns, numbers of which have been poetically-named by the Chinese, as can be seen from inscriptions carved in the stone overhead. The cavern interiors showcase the fine sculptural phenomena peculiar to limestone eroded by acidic water.

AYERS ROCK

The red giant at the heart of the outback

Merely a blacker bulk against the desert night, Ayers Rock begins to lighten as the sun spreads its dawn rays across the sky. Shifting from black to deep mauve, the gigantic monolith gradually becomes more distinct. When the first rays of the sun actually strike, the stone bursts into a riot of reds and pinks that chase each other across the surface with startling speed. Shadows flee from lumps and hollows until the whole rock is bathed in the harsh light of the desert. The colour changes continue more sedately throughout the day; come sunset and the stone again races through its fantastic spectrum.

Approaching Ayers Rock along the bumpy and often unsurfaced road from Alice Springs, it is the sheer bulk which impresses. The rock rises dramatically above a vast plain without the warning of lesser hills, and is given scale only by sharply-pointed spinifex grasses and desert oaks. Some 3km (2mi) long and 348m (1,142ft) high, it sits on the horizon, looking like a stranded whale. But the rock is, in fact, more akin to an iceberg, since much of its bulk lies below ground – some geologists estimate the rock reaches 6km (3.75mi) down into the Earth's crust. And around its perimeter, a distance of more than 8km (5mi), are numerous caves and hollows sacred to the Aborigines.

A solid block of sandstone conglomerate, the rock is one of the few remnants of a primeval ocean floor that occupied the centre of Australia around 500 million years ago. Gradual upheavals and movements of the Earth's crust have upturned the rock's formerly horizontal layers; weathering has fashioned their exposed edges into ridges and grooves. For hundreds of thousands of years the rock has defied the forces of erosion while the red centre of Australia has changed from a lush and fertile landscape to a desert. Whenever the rains do fall, the mighty rock

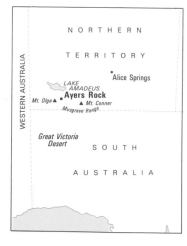

Ayers Rock protrudes from a flat plain around 320km (200mi) to the southwest of Alice Springs in Australia's Northern Territory (*left*). Visible from as far away as 96km (60mi), the rock (*right*) is a solid lump of arkose, a sandstone rich in feldspar. This iron ore is responsible for the rock's golden red hue, and for the dramatic colour changes witnessed at dawn and dusk.

takes on yet another hue – the liquid silver of a skin of water. Surface grooves become raging torrents. Water cascades down to form pools on the desert floor, triggering shield shrimps to wake up from their dormancy, an act these tiny creatures and their predecessors have performed unchanged over the last 150 million years or more.

Sacred aboriginal site

At close range the rock reveals an incredible delicacy of form sculpted by falling waters and howling winds. The Brain, for example, is a shallow depression ridged and furrowed just as a brain is. A smoother cavity is the Sound Shell, which lives up to its name when the wind blows. A cave at the base where erosion has scooped out a long narrow hollow from the sandstone is shaped like a huge breaking wave. Dwarfing the humans who cluster around its foot, this petrified wave seems set to crash to the ground at any moment.

At the western end of the rock, iron spikes driven into the stone support an iron chain that climbs up the steep slope and disappears from view. This is the safe route to the top. A plaque at the bottom of the chain recalls the tragic end of those who tried to find other routes. The climb, a distance of 1,612m (5,289ft), is lung-bursting and tiring. But from the cairn on the summit a splendid view across the baked desert is just reward. To the west lie the Olga Mountains; to the south, the Musgrave Ranges; and to the east, the flat top of Mount Conner.

The massive bulk of Ayers Rock is called Uluru by the local tribes of Aborigines, the Yankunitjatjara and the Pitjantjatjara. Many sacred paintings relating to the Dreamtime decorate their caves which tourists are forbidden to enter. To the Aborigines the Dreamtime was the beginning of all things. It was the time when gods and animal-men roamed the country singing every feature of the landscape into existence – rocks, mountains, streams – while creating the sun and founding tribes. Every individual feature of Ayers Rock, whether it be a crack, an indentation or a lump, has a meaning to the local Aborigines.

The neglected Olga Mountains

English-born explorer Ernest Giles (1835–1897) was the first white man to set eyes on Ayers Rock. He sighted it in 1872 from the boggy shores of Lake Amadeus, 40km (25mi) to the north, on one of several expeditions he made to the Australian interior. Forced to abandon his journey, Giles returned the following year, only to find that an expedition from the surveyor-general's department led by William Gosse (1842–1881) had climbed the rock before him. Gosse named the isolated monolith after Sir Henry Ayers (1821–1897), who was at that time Premier of South Australia.

Giles became spellbound by the beauty of the Olgas – a group of 30 giant rocks 32km (20mi) to the west of Ayers Rock. The tallest of these, Mount Olga, rises 546m (1,791ft) above the sandy plain. He named these rocks after the Russian princess who had married the King of Spain, and wrote that: 'Mount Olga is the more wonderful and grotesque; Mount Ayers the more ancient and sublime.'

The local Aboriginal name, Katatjuta, means 'the place of many heads' for the Olgas are tall and rounded like the crowns of giant heads buried in the sand. They are almost as spectacular as Ayers Rock but receive far less attention. Perhaps this is because the Olgas, while composed of the same hard sandstone, lack the abundance of iron ores which cause Ayers Rock to change colour so dramatically in the sun.

A gigantic rib of rock known as the Kangaroo's Tail leans against the curving north side of Ayers Rock. When exposed to the alternate extremes of heat and cold, which are prevalent in the desert, the monolith's sedimentary sandstone expands and contracts. Eventually, in a process known as 'spalling', the sandstone cracks and sheds thin flakes of rock which usually tumble and smash to the ground. But a few, such as the Kangaroo's Tail, remain intact. Spalling takes place at roughly the same rate over all parts of Ayers Rock. As a result, the rock is gradually shrinking in size, but without changing its overall, distinctive shape.

Persistent weathering of Ayers Rock causes its grooved surface to become honeycombed with caves, such as the hollow. This can be found beside the place known by Aborigines as the Camp of the Sleepy Lizard. The caves and hollows are etched by rainwater, which loosens the sand grains and pebbles binding the rock's sandstone together. On the north side of Ayers Rock, an assembly of hollows has created a convoluted and furrowed honeycomb appropriately entitled the Brain.

A terrifying devil dingo once threatened the Pitjantjatjara, the aboriginal hare-wallaby people whose Dreamtime home was on the north side of Ayers Rock. The tribe survived only by escaping in great leaps and by snatching the beast's totem from its mouth. Yet it was the Aborigines who introduced dingos (*Canis dingo*) to Australia thousands of years ago. Largely indistinguishable from domestic dogs, dingos breed in burrows or rock crevices and prey on sheep, rabbits and kangaroos.

MOUNT FUJI

Sacred peak of perfection

The perfect symmetry of Mount Fuji's silhouette has long been the ultimate Japanese symbol of beauty. This revered volcanic peak stands snow-clad for much of the year, while its lower slopes lie cloaked in lush vegetation or burnished by moorland. English traveller and broadcaster John Morris (1895–1980), who visited Mount Fuji immediately before the outbreak of World War II, declared the peak was best seen from the sea at dawn: 'At such times it seems to tower over everything, its perfect snow-capped cone, a purplish green in the light of early morning, seeming to be suspended in the sky.'

The aboriginal Ainu people revered the mountain centuries before the Japanese colonized the country around 2,000 years ago. Ainu tribes still survive on Hokkaido, Sakhalin and other islands in the Pacific Ocean to the north of Japan. It was they who named the peak 'Fuji', which has been roughly translated as 'everlasting life' or, alternatively, as 'fire goddess'. The Japanese kept the name and maintained the mountain's holy tradition.

According to Buddhist teachings, which reached Japan in about AD 550, Fuji was created when a massive earthquake struck the land late one night in 286 BC. The same seismic upheavals opened up the earth and formed Lake Biwa, Japan's largest body of inland water, some 280km (175mi) to the west.

The volcano that is Mount Fuji first erupted around 300,000 years ago from below a wide plain. Outpourings from several cones have helped fashion Fuji's present shape, by building up alternate layers of solidified lava and a conglomerate composed of cinder, ash and lava. These layers represent the sequence in which the volcano, known to geologists as a strato-volcano, erupts: huge volumes of molten lava spread out evenly over the mountain slopes. Violent explosions follow, in which dense

Mount Fuji stands in the south-central region of Japan's chief island of Honshu, some 100km (62mi) southwest of Tokyo, the country's capital (*left*). Known also as Fuji-san and Fujiyama, the sacred peak lies in the Fuji-Hakone-Izu National Park, an area of 1,222sq.km (472sq.mi) established in 1936. The most perfect of all the views of the peak is from across Lake Kawaguchi (*right*) when the mountain's symmetrical cone is reflected on the water's surface.

clouds of cinder, ash and lava pellets are ejected high into the air.

Since Mount Fuji's first recorded eruption in AD 800, the volcano has discharged lava on ten recorded occasions. Each time, the outpourings concealed the remnants of two ancient craters, known as Old Fuji and Komi-Take. Clouds of ash and cinder from Fuji's last eruption in 1707 were carried as far as Tokyo, 100km (62mi) to the east, where streets were blocked and buildings damaged.

The currently dormant volcano is Japan's highest peak, rising to a height of 3,776m (12,388ft). From its crater, which measures almost 700m (2,297ft) across, Fuji slopes down at an angle of 45 degrees and gradually levels out before reaching the plain. The mountain's base traces a near-perfect circle, 40km (25mi) in diameter and 125km (78mi) in circumference.

Climbing the symbolic mountain

As the holy mountain of Shinto (the official religion of Japan), Fuji was forbidden to women until 1868. The first woman recorded as having attained the summit, the wife of the British Ambassador to Japan, had in fact completed the ascent the previous year. Until the end of World War II, the summit had been the goal of every devout follower of Shinto. Climbers were expected to carry a rock to lighten the souls of sinful people.

Since World War II, the sacred mountain has lost an increasing amount of spiritual ground to the profane desires of tourism. By the 1980s, as many as 300,000 sightseers made the arduous climb each year. This figure seems the more remarkable because the five paths to the summit are open only in July and August – for the remainder of the year the mountain is cloaked in snow. Stations along the five paths offer beds and refreshment to pilgrims and tourists aiming to view the *goraiko*, the unique sunrise which can be witnessed from Fuji's summit. The strenuous and unique nature of the journey is reflected by the Japanese proverb: 'It is as foolish not to climb Fuji-san as to climb it twice in a lifetime'.

John Morris, like every other visitor to reach the summit, was deeply moved: 'The view from the top was astounding: over a maze of still, dark lakes and valleys we looked out right across the Pacific, the line of the coast faintly visible, like a meandering smear brushed in with purple ink.' The *goraiko* is a special sunrise because, immediately before the sun appears above the horizon, its disk is reflected upon the ceiling layers of the atmosphere, creating a wash of colours that vanishes with the first direct rays of light.

Mount Fuji is worshipped as the abode of the gods, the symbolic link between the mysteries of heaven and the realities of everyday life. More than any other feature of the nation's culture, Mount Fuji is the emblem of Japan. For 12 centuries, the country's poets, painters and writers have attempted to capture its inherent beauty and capricious moods. The '36 views of Mount Fuji' by the painter Katsushika Hokusai (1760–1849) have become the archetypal pictorial representations of the peak.

During the 13th and 14th centuries, Mount Fuji played an important part in the development of zoen, the art of landscaping. This art was influenced by the thoughts of Zen Buddhism, which emphasized contemplation of the essence of nature. The idea was to create a scene which, when viewed from a house or vantage point, showed no signs of human interference and blended together into a harmonious whole. Around Mount Fuji, most of the landscape artists have used the peak as a distant focus. The quintessential zoen landscape can be witnessed from Mount Tenjo to the north when, on clear days, Mount Fuji is fully reflected in the still waters of Lake Kawaguchi below.

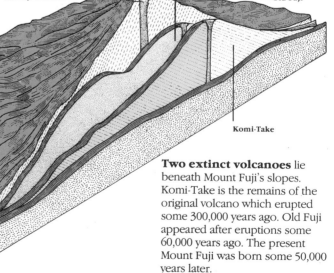

Two extinct volcanoes lie beneath Mount Fuji's slopes. Komi-Take is the remains of the original volcano which erupted some 300,000 years ago. Old Fuji appeared after eruptions some 60,000 years ago. The present Mount Fuji was born some 50,000 years later.

The apex of Mount Fuji soars above the countryside of Japan like a rare and sacred flower. Mount Fuji is the tallest of Japan's 'mountains of fire' and arguably the most perfectly-shaped of all volcanoes. Its crater forms an immense orifice, some 503m (1,650ft) wide and 221m (726ft) deep. The volcano's summit is snow-clad for 10 months of the year – only the rocky crags perched on Fuji's peak remain snowless due to high winds. Eight crests ring the brow of the crater: these are known as the Yatsudo Fuyo, or 'the eight petals of Fuji'. The highest crest, Kengamine, stands 76m (250ft) above the crater's rim.

Mount Fuji has been revered for centuries by the people of Japan as a 'fire goddess' or the 'abode of the gods'. Many religious zealots have treated the mountain as a sacred shrine: climbing it became an essential step in the development of a pilgrim's soul.

The most fervent worshippers are the Fujiko, a religious sect founded in 1558 by Takematsu, who climbed the mountain 120 times. From the summit, pilgrims watch the unique sunrise, known as the *goraiko*, or meditate on the tranquillity of the lakes below, such as Lake Yamanaka (*right*).

GREAT BARRIER REEF

Living coral gardens at the edge of the Pacific

The 18th-century English explorer Captain James Cook discovered the Great Barrier Reef by the simple, if dangerous, method of sailing straight into it. While seeking the great Unknown Southern Land rumoured to exist in the Pacific Ocean, Cook inadvertently stumbled upon one of the Earth's greatest and most fascinating underwater treasures.

After completing astronomical observations near Tahiti, and having charted the coast of New Zealand, Cook headed west. On April 19, 1770, he sighted the coast of the land that was later to be called Australia. In response, Cook turned north and entered treacherous coral waters. He soon realized he was sailing through the calm, shallow lagoon waters usually found between a reef and a coast. But despite Cook's precautions to avoid the reef his ship *Endeavour* became stuck fast on an isolated pinnacle of coral.

Cook recorded in his journal: 'A Reef such as is here spoke of is scarcely known in Europe . . . it is a wall of Coral Rock rising all most perpendicular out of the unfathomable Ocean . . . the large waves of the vast Ocean meeting with so sudden a resistance make a most terrible surf breaking mountains high especially as in our case when the general trade wind blowes directly upon it.'

Cook's men spent two months repairing *Endeavour* but were still hemmed in by the reef. Not until they reached the point now called Cape Melville did they escape the coral maze and enter the open sea, by a route known today as Cook's Passage. Later, while heading northward, they were in danger of being smashed by perilous swells against the coral walls. Once again Cook searched desperately for a break in the reef. Just in time he found a narrow opening into the tranquil lagoon waters which consequently he named Providential Channel.

The Great Barrier Reef stretches for more than 2,010km (1,250mi) along the northeastern coast of Australia (*left*). From off Bundaberg to Papua New Guinea, the whole reef complex covers an area of 259,000sq.km (100,000 sq.mi), roughly a third of the size of the state of New South Wales. Rich in colour and form, the reef's underwater gardens are composed of myriad corals, such as the table coral (*Acropora hyacinthus*) (*right*).

Along the eastern seaboard of Queensland a wide continental shelf of limestone rock reaches out to the Pacific Ocean. The clear, salty, well-oxygenated waters above it provide an ideal breeding ground for a rich variety of hard or stony corals. But these creatures will only colonize into reefs if the water maintains a year-round temperature not less than 20°C (68°F).

Simple animals related to the more familiar sea anemones, hard corals average about 5mm (0.2in) in diameter. Their white skeletons, rich in calcium carbonate, are clothed in soft, variously shaped and brightly coloured tissue. These animals contain single-celled algae called zooxanthellae which provide the vital metabolic support they need for secreting calcium carbonate and, hence, for reef building. The presence of these algae means that coral reefs will only form in shallow water, since below 55m (180ft) the algae are unable to acquire the light they need for photosynthesis.

More than 350 species of coral thrive on the hard limestone bedrock formed from the skeletons of their countless ancestors. The vast, kaleidoscopic colonies they build are homes to more than 1,400 species of fish, myriad sponges and numerous echinoderms, such as sea urchins and brittle stars. Echinoderms called sea cucumbers play an important role in consolidating the reef's structure. As detritus feeders they excrete tiny fragments of shell and sand which sink down to the depths and compound the reef's foundations.

Despite the exacting environmental requirements of healthy coral, the Great Barrier Reef has a total length of around 2,010km (1,250mi). In its southern reaches, near Cape Manifold, both the continental shelf and the reef are at their widest – around 320km (200mi). Here, where the shelf plummets to the ocean floor, the turquoise, shallow waters contrast brilliantly with the Pacific's deep blue. Farther north, off Cape Melville, the reef is merely a narrow band of surf-fringed coral hugging the coast.

Ninety per cent of the reef lies underwater. Breaking through the uncommonly clear sea, an archipelago of islands, shoals and reefs is separated by twisting channels, which only small craft can navigate with safety. In other places the coral is so extensive that people can wade across what appears to be huge stretches of open water without ever needing to swim.

The starfish menace

The reef's variety of colours and shapes never ceases to astonish. The multiple branches, giant fans or iridescent tentacles of the corals exist in concert with the dazzling patterns and vivid adornments of fish, crustacea and worms. In every direction, brilliantly patterned creatures dart in and out of coral gardens or lie in wait concealed in bizarre camouflage.

Yet this glamorous spectacle of Nature may be in danger. Apart from the hazards of imminent oil exploration, the corals are threatened by one of their own inhabitants, the predatory crown-of-thorns starfish (*Acanthaster planci*). This animal appeared on the reef in large numbers in the 1960s. By the 1970s, huge areas of the reef had been stripped of their living coral, leaving behind only bare limestone. A single starfish may be up to 40cm (16in) across with a maximum consumption of around 100sq.cm (16sq.in) of coral a day.

Since the 1970s, starfish numbers have been declining, so allowing the coral to regenerate itself – although this could take 30 or 40 years to be effected completely. Scientists studying the reef have found evidence of similar devastations in the past, which suggests that this menace, like much else in Nature and the reef, is part of the normal cycle of events.

When Captain James Cook (1728–1779) was aged 40, the British Government gave him command of an expedition to chart areas of the South Pacific Ocean. In his 368-tonne ship *Endeavour*, the self-taught navigator carried out explorations that revolutionized knowledge of the southern seas. It was on this voyage, the first of three he would make to the region, that Cook discovered the Great Barrier Reef.

A reef walk out across the coral platform on Heron Island at low tide allows close range inspection of many colourful and fascinating species. Here, fish, corals, molluscs, echinoderms all inhabit sunlit tidepools. Cobalt-blue seastars, sausage-like black sea cucumbers, sea anemones and giant clams with brilliant gem-green ruffled mantles are commonplace.

Heron Island, 70km (44mi) off the Queensland coast, is rare in being a genuine coral cay – a part of the reef itself – as opposed to the many continental islands in the reef area.

North Reef, almost circular in shape and a scant 2km (1.25mi) across, lies along the Queensland coast at the southern end of the Great Barrier Reef. Its most important feature is a marine light, alerting ships to the fact that this is where the Capricorn Channel suddenly alters its width – narrow to the south and broad to the north.

Clown Triggerfish

Clown Anemonefish

Sweetlip Emperor

The reef is home to more than 1,400 species of fish, among them: the clown triggerfish (*Balistoides conspicillum*), a flamboyantly patterned individual which enlarges itself in defense by expanding a flap on its belly; the clown anemone fish (*Amphiprion percula*), which shelters from predators by swimming inside large sea anemones – its own body mucus protects it from the anemones' poison; and the sweetlip emperor (*Lethrinus chrysostomus*), a tasty perchlike fish also known as the 'tricky snapper' which weighs as much as 9kg (20lb).

ROSS ICE SHELF

The biggest iceberg in the world

'Well, there's no more chance of sailing through that than through the cliffs of Dover', exclaimed Captain James Clark Ross in 1841 when he first saw the spectacular ice shelf that would later bear his name. Ross had been instructed by the British Government to explore the southern regions of the globe. Most importantly, he was to reach the South Magnetic Pole.

The reports of earlier British explorers to the Antarctic, especially James Cook, who had sailed there in 1773, and James Weddell, who reached it in 1823, had been encouraging. Although these men had encountered seas full of icebergs, they had not sighted land. In 1831, following similarly encouraging reports from the Arctic, Ross's uncle, John, had reached the North Magnetic Pole.

In November 1840, the expedition's two small, sturdy ships, the *Terror* and the *Erebus,* sailed into the iceberg region of Antarctica. For two months, Ross nosed his ships slowly through the treacherous water. When his lookout uttered a cry of triumph, Ross saw through his telescope a stretch of clear ocean running southward to the horizon. Was it possible he would have a trouble-free journey to his destination?

The expedition's high hopes were dashed when they saw the enormous shelf of ice loom before them, preventing them from navigating farther. For two years, Ross steered his ships around the coast of Antarctica, but found no sea passage to the South Magnetic Pole. Though Ross had failed in his prime objective, his detailed journal, complete with charts and illustrations, would prove indispensable to later explorers in the region.

As the world's largest body of floating ice, the Ross Ice Shelf has staggering dimensions. The massive slab of ice is around 800km (500mi) long and all but fills a huge bay of the Antarctic

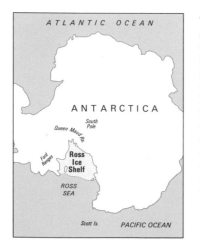

The Ross Ice Shelf extends some 965km (600mi) inland from the Ross Sea toward the South Pole (*left*). The surface area of its ice covers around 520,000sq.km (200,000sq.mi), almost the size of France or the state of Texas. The formidable blue-green cliffs at the seaward edge of the shelf (*right*) are like the end of an immense conveyor belt: mammoth icebergs as long as 40km (25mi) are regularly sheered off and cast into the currents of the open sea.

continent. In the shelf's southern reaches, nearest the true pole, the ice can be as thick as 750m (2,400ft), while in the north it is roughly half this figure.

The seaward side of the ice shelf is forbidding in the extreme. When seen from afar it appears to be the straightest line in nature. At close quarters, the ice cliffs can be seen to rise as much as 70m (230ft) above the wind-shredded sea, and sparkle an ethereal blue-green in the sun. Moreover, the ice shelf moves. With remorseless power the huge sheet of ice edges toward the sea at a rate between 1.6m (5ft) and 3m (10ft) each day. This constant motion is generated from three sources: glaciers from behind, snow from above and ice from below.

Several huge glaciers, such as the Beardmore Glacier, flow down from the distant Transantarctic Mountains. They add continuous quantities of ice to the rear of the shelf, nudging it forward. The enormous amounts of snow that fall on the shelf's surface each year are unable to melt because of the cold air. Layer upon layer of fresh snow builds up until the weight of the upper layers compacts those below into ice. Simultaneously, the freezing sea beneath the shelf pushes upward. These constant additions to its weight cause the lower layer of the shelf to flatten out and move in the only direction possible – to the sea.

Crevasses and fractures in the shelf can run for many miles, often forming inlets of open water. When such a fracture encircles a mass of ice, gigantic blocks break free from the main shelf. Drifting as vast tablelands of ice, such bergs are commonly around 40km (25mi) long. In 1956, an iceberg with fantastic dimensions was spotted off the coast of Scott Island, some 1,280km (800mi) north of the Ross Ice Shelf. Covering an area of around 31,000sq.km (12,000sq.mi) and with a maximum length of 335km (208mi), it was larger than Belgium.

The race for the South Pole
The unwelcoming sight of the ice cliffs contrast with the upper surface of the shelf, which is remarkably level and firm like a plain. Because of its flatness, the shelf has been the starting point and base camp for numerous expeditions within Antarctica. In 1908, British explorer Ernest Shackleton landed at the shelf's western end with an expedition determined to be the first to reach the South Pole. Before they made the final assault, three members of his party achieved what Ross had sought and established the position of the South Magnetic Pole.

On October 29, Shackleton set out with four pony-drawn sleds and climbed the Beardmore Glacier to reach the polar ice cap. His party was 160km (100mi) short of the pole when the death of the ponies and the lack of food forced them back. Yet he had blazed a trail which others would soon follow.

In January 1911, Captain Robert Scott landed on the Ross Ice Shelf, close to Shackleton's old base. At the same time, Norwegian explorer Roald Amundsen landed to the east. The race for the pole was on. The two expeditions set out when the spring brought longer hours of sunlight and good weather. Because he chose a smoother route and had stronger dogs, Amundsen reached the pole first, on December 14, 1911. Scott's team took a month longer.

On the return journey, Scott and his companions suffered terrible hardships when their animals died. At the edge of the Ross Ice Shelf, Captain Titus Oates slipped away to die of frostbite rather than slow down his friends and endanger their lives. But, a few days later, the whole party perished of exposure and lack of food. In Scott's diary, rescuers found the telling words: 'Great God! This is an awful place.'

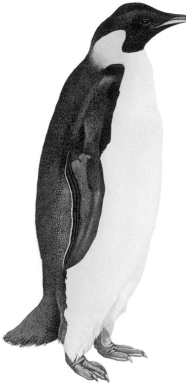

The emperor penguin
(*Aptenodytes forsteri*) is the only animal to endure the antarctic winter. When all other creatures have fled, the emperors return from the sea to breed in the most hostile conditions known to any bird. Assembling on the ice shelf beneath the cliffs in their thousands, they pair off. Mates identify one another by means of each bird's unique song. No nesting can be done in this frozen habitat, so the male carefully incubates the single egg laid by the female, resting it on his feet and settling his warm belly over it. When the chick hatches, both parents feed and rear it.

When a sailor's-eye view of formidable icebergs first greeted Captain James Ross in 1841, he reckoned the ice shelf's frozen escarpment was impassable. And so it was, by ship. However, the flat upper surface was to prove a practicable base for later explorers, and in modern times, aircraft, motorized transport and icebreaking ships continue to improve access to the continent.

Antarctica now has dozens of permanent research stations, such as the one at McMurdo Sound at Ross Ice Shelf's eastern edge. There scientists conduct a range of experiments, and investigate subjects including meteorology and the changing composition of the atmosphere's endangered ozone layer.

Captain Robert Scott and his comrades reached the South Pole in mid-January, 1912, only to discover that they had lost the race to rival explorer Roald Amundsen. The small flag flown over Amundsen's abandoned tent proclaimed a Norwegian victory (by 35 days) to the bleak, surrounding icescape. All of Scott's five-man party perished on the return trek and the film containing this photograph was recovered from their effects.

Amundsen's successful dash was begun from the Bay of Whales, an inlet of the Ross Sea at the ice shelf's eastern end. His speedy crossing of the shelf to his goal owed much to the Norwegian skill of cross-country skiing. He survived Scott by 16 years, succumbing himself in the Arctic while attempting to rescue Italian pilot Umberto Nobile from a crash in the ice.

MILFORD SOUND

Zoological zones in fjord waters

The majestic beauty of Milford Sound's towering cliffs, dense forests and glittering waters prompted English author Rudyard Kipling (1865–1936) to declare it 'the eighth wonder of the world'. The most northerly of the dozen or so fjords which bite deep into the southwest corner of New Zealand's South Island, Milford Sound is more than an inlet of scenic magnificence. Within its waters and surrounding terrain dwells a unique collection of flora and fauna.

Enormous glaciers from the mountains fashioned the valley of Milford Sound around 20,000 years ago. When, some 10,000 years later, the glaciers retreated, the waters of the Tasman Sea rushed in to create a fjord 19km (12mi) long and 2.5km (1.5mi) at its widest. Moreover, the world's highest sea cliffs rise up almost 1.6km (1mi) from the water's edge and descend beneath the water to depths of 396m (1,300ft). The retreat of smaller, tributary glaciers created 'hanging valleys' in which streams spill their waters into Milford Sound as waterfalls with drops of more than 300m (1,000ft).

At the entrance to the fjord stands the awesome Mitre Peak which rises in an almost-perfect pyramid to 1,695m (5,560ft) above the sea. Its slopes, and those of other mountains surrounding the fjord, are blanketed with thick forests, most commonly of beech trees. These woodlands, which are protected by the Fjordlands National Park, shelter the rare and peculiar bird known as the takehe (*Notornis mantelli*). This flightless bird, clothed in purple and blue plumage, is about the size of a chicken. It was thought to be extinct until, in 1947, a colony of about 100 birds was discovered in a side valley off Milford Sound. Almost as rare is the kakapo (*Strigops habroptilus*), a ground-dwelling parrot that resembles an owl and lives in a burrow by day.

Milford Sound is located in the southwest corner of New Zealand's South Island (*left*). It is an isolated inlet, flanked by the world's tallest sea cliffs, and the majestic climax to the Milford Track. This undulating path, which winds over mountains and through forested valleys from Lake Te Anau, some 55km (33mi) to the southeast, has been called 'the finest walk in the world'. Mitre Peak (*right*) stands like a giant sentinel over the dark waters of the sound, and is the highest point on the surrounding mountains.

The water in Milford Sound has an unusual composition. Two factors are instrumental in defining the layered nature of the water. First, the mouth of the fjord, which is both narrow and shallow, restricts the inward and outward flow of seawater. This was because the glacier lost its power before reaching the sea, leaving intact the rock walls at the seaward end of the valley. It also left a ridge, or sill, of rock across the entrance which means that the water here is only 55m (180ft) deep.

The second factor is the prodigious rainfall of the region. The average annual precipitation at Milford Sound is 6.43m (21.5ft). This enormous volume of water is deposited over the surrounding mountains and percolates through the deep beds of moss and humus-rich soils which carpet the forested slopes. By the time the water reaches the fjord, via myriad mountain waterfalls and streams, it is stained the colour of tea by the organic materials it has collected. When this rainwater arrives at the fjord it does not mix with the saltwater but, being less dense, forms a layer 3m (10ft) thick on top.

This freshwater film gradually moves seaward, dragging with it the upper layers of saltwater. This movement creates a counter-current and draws seawater into the fjord to a depth of 30m (100ft). Because of this superficial exchange of seawater and freshwater, much of Milford Sound's deep water has become almost deoxygenated and stagnant.

Marine life on the fjord's walls

Scientists led by Ken Grange of the New Zealand Oceanographic Institute (NZOI) concluded that this layer of freshwater has had a profound effect on the marine life of Milford Sound. Most noticeable is the fact that the tidal waters are free of the algae and molluscs commonly found on the coasts of New Zealand. The reason is because they cannot tolerate the low levels of salinity. Instead, the intertidal zone on the fjord walls is colonized by species of snail, green sea lettuce, blue mussels and barnacles, all of which are normally associated with brackish waters.

In the 1980s, research by NZOI showed that the flora and fauna that do survive here, especially on the rock walls below the intertidal zone at depths from 6m (20ft) to 40m (131ft), are those normally associated with deeper seawater. The reason is the dark colour of the top layer of freshwater, which restricts the penetration of sunlight and encourages species from deeper waters to occupy niches higher up among the ledges and fissures of the steep, rocky walls. Such species include sea pens, sea feathers, sea squirts, gorgonians and brachiopods.

The most intriguing species to be discovered in this intermediate zone is the black coral (*Antipathes aperta*). While they are alive the polyps of these corals are, in fact, white; their bodies blacken only when they die. The species usually flourishes in colonies below depths of 45m (148ft), but in the fjord the black coral grows at less than 35m (115ft). Most colonies are feathery bushes no more than 10cm (4in) tall, but a few are 150-year-old trees as tall as 4m (13ft).

Scientists at NZOI estimate that there are approximately 7.5 million colonies of black coral growing in Milford Sound and the other fjords on New Zealand's South Island. This represents the largest known resource of black coral in the world. Because such colonies are within easy reach of scuba divers, they are in grave danger of being harvested – its attractive shape and captivating colour have already made black coral a favourite material for rings and brooches elsewhere in the Pacific. This would be disastrous since black coral. which grows on average at a rate of 2.5cm (1in) per year, takes many years to regenerate.

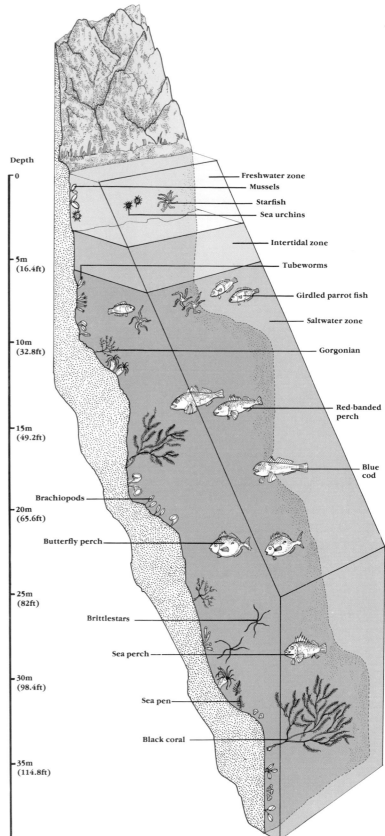

The unique ecology of Milford Sound's stratified marine habitat is brought about by the combination of several physical factors. Steep mountain walls inhibit light yet provide shelter. Below the dark-coloured freshwater, the saltwater is unusually still, clear and warm. Annual water temperatures are 11°–15°C (52°–59°F): this is the smallest range for seawater found anywhere on New Zealand's extensive coast.

Hanging Veil Falls is typical of the spectacular cascades seen everywhere in Fjordland National Park. All around Milford Sound – itself a glacier-carved fjord – water is literally left hanging by the retreat of ice-age glaciers. This plummets down the mountainsides in cascades, which produce a veil-like spray as the water atomizes on impact with the rocks below.

Along the road from Lake Te Anau to Milford, inland from the fjord's edge, countless ribbon-fine streamers of white pour down the near-vertical faces of jet-black peaks. Below lie chaotic scree slopes of tumbled boulders, as eerie as a building site on another planet.

From the Milford Track (the trekker's alternative to the road) can be seen the Sutherland Falls. At 580m (1,904ft) they are the highest in New Zealand, and among the highest in the world.

WHAKAREWAREWA

Hot springs and geothermal fountains

In March 1886, a fully equipped Maori war canoe was seen speeding across a lake in the centre of New Zealand's North Island. The local newspapers pondered that the country was at peace; such canoes had not been seen for many years. The local Maori, declaring the vessel to be an old ghost which had come to predict disaster, waited for something to happen. They did not have to wait long.

On June 10 of that year, the volcanic peaks of Mount Tarawera near Rotorua blasted themselves apart in a series of massive explosions. Lava gushed forth and lakes boiled dry. The earth shook and rocked. The eruptions ceased almost as quickly as they had begun and left, in the midst of a reshaped landscape, one of the world's most dramatic geothermal regions. Here, in an area of only a few acres, turbulent hot springs and spectacular geysers continue to emit their sulphurous water.

'I thought it an uncommonly pleasant place, although it smells like Hades,' commented the Irish playwright George Bernard Shaw (1856–1950) after a tour of New Zealand's volcanic plateau in 1934. He found Tikitere, 16km (10mi) to the east of Rotorua, to be especially infernal: 'Tikitere, I think, is the most damnable place I have ever visited and I would willingly have paid ten pounds not see it.'

Whakarewarewa is only a short walk from the centre of Rotorua, its world-famous thermal attractions adjacent to a traditional Maori village. In boiling mud pools, popularly known as 'porridge pots', glutinous mud shifts and swirls in endlessly changing patterns. Bubbles of sulphurous gas find their way to the surface, exploding with loud plopping sounds and sending tiny globules of hot mud into the air as they do so.

Terraces composed of crystalline silica provide the backdrop

Whakarewarewa is a Maori suburb of Rotorua, the chief city of New Zealand's volcanic plateau (*left*). Amid boiling mudpools, steaming fumeroles and hissing vents, the country's tallest geyser, Pohutu, reaches for the sky (*right*). One of seven geysers to erupt in this spectacular wonderland, Pohutu has twin jets of steam and water which often climb to a height of 30m (100ft).

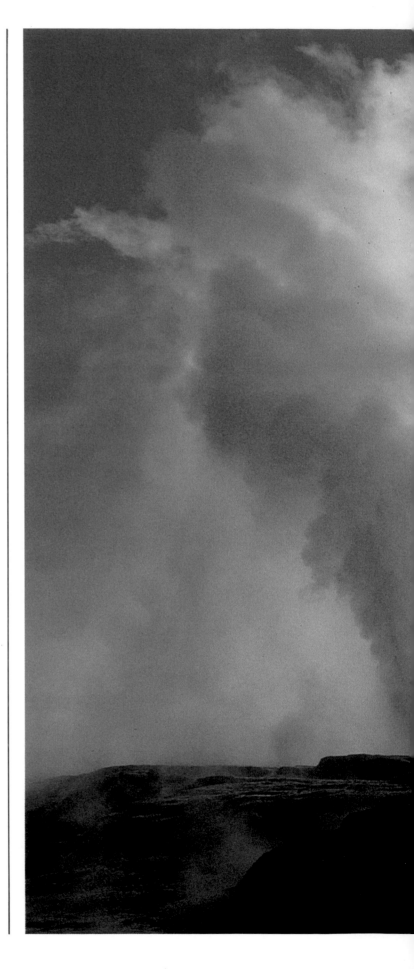

for Geyser Flat, where Whakarewarewa's seven geysers can be seen on display. For much of the time, they are quiet pools of steaming water. But at frequent, though irregular, intervals they erupt into life, rolling and bubbling before sending fountains of water high into the air.

Pohutu is the largest of these geysers. In fact it is the largest geyser in New Zealand – its dramatic plume of boiling water and steam can be as tall as 30m (100ft) high. By comparison, the water jets of Strokkur in southwest Iceland reach about the same height; those of Old Faithful in Yellowstone National Park, Wyoming, USA, gush to 40m (130ft). Pohutu, from the Maori word meaning 'splashing waters', is notoriously impossible to anticipate. Months may pass before it stirs, but when it does the great geyser may erupt several times a day.

Pohutu's activity is usually preceded by a smaller geyser, the Prince of Wales Feathers, whose waters emerge through three outlets and create a display resembling the ostrich-feather emblem of Britain's Prince of Wales. Both geysers draw on the same reserve of underground water and erupt in a predictable pattern: the Prince of Wales Feathers plays for between two and five hours, sending its jets of water to a height of 12m (40ft). Then, as soon as it dies down, the majestic Pohutu bursts forth.

The natural spectacles of geysers, blow holes and mud pools alerted engineers to a hidden source of usable power within the Earth. In 1961, at Wairakei Valley near Taupo, New Zealand's first geothermal power station went on line. This was also the world's first hot-water, or wet-steam, geothermal power station. Engineers tapped the Earth's heat, or geothermal energy, harnessed it to huge steam turbines and generated electricity. Cheap and environmentally clean – there is none of the pollution associated with oil- or coal-fired power stations – geothermal energy could provide New Zealand's North Island with more than half its electricity needs.

The springs of Waimangu

To the east of Whakarewarewa are the uncanny thermal springs at Waimangu, the site of what was once said to be the world's most spectacular geyser. Its first recorded eruption was in 1900. In 1904, it sent up a stream of scalding water which reached an estimated 450m (1,500ft). Over the succeeding years its activity became less frequent and reduced in size until 1917 when it ceased altogether.

Most of the Waimangu Hot Springs are small gushings of water but a few, such as the Waimangu Cauldron, are awe-inspiring in their intensity. Rising above one side of this lake are the Cathedral Rocks. Clothed in lush vegetation, they steam constantly as boiling water emerges through cracks in the rocks and cascades to the cauldron below. On cold days, the lake is blanketed by an eerie mist of water vapour which light breezes and air currents twist into strands and move around in a strange, unnerving manner.

Lake Rotomahana was the main attraction in this region until the land around Waimangu was drastically altered by the great eruption of Tarawera. This catastrophe increased the depth of the lake 20-fold to 213m (700ft), drowning the renowned Pink and White Terraces around its waters. Built up over centuries by the accumulation of minerals in the hot spring water, these terraces covered almost 5 hectares (12 acres) and sparkled in many graded shades of white and pink. When the lake's green and blue water trickled over them they glistened in a rich array of delicate pastel hues. Unless the waters of Lake Rotomahana recede to their former levels, these tinted silica terraces will be lost to the world forever.

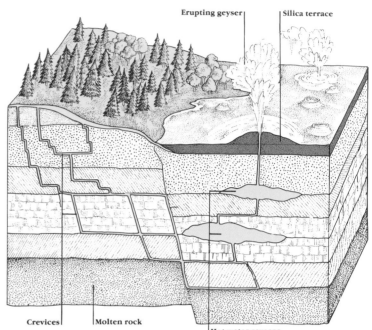

Erupting geyser | Silica terrace

Crevices | Molten rock | Hot water caverns

Geysers are characteristic of geologically 'young' areas, such as New Zealand, where molten rock known as magma flows near to the Earth's surface. Seepage from crevices in the ground gathers in subterranean caverns in the hot rock layer overlying the magma.

The accumulated water, heated to boiling point, must then find an escape route upward. Steam pressure causes it to burst forth in the form of a geyser, which plays for a time and then subsides until the pressure builds up again. Geysers have individual rhythms of performance, but even these are subject to variation.

Sparkling silica terraces were built up from various minerals dissolved underground by the geysers' vaporized hot waters. Shrouded in steam, the terraces are pocked with the seven pools that erupt into Whakarewarewa's world-famous fountains.

Bubbling, spitting fumaroles emit the sulphurous gases (*left*) that give Whakarewarewa its unenviable smell of rotting eggs. The boiling mud pools are nicknamed 'porridge pots', an apt description of their gluey consistency.

Whakarewarewa's Boxer geyser punches the sky with a shower of shimmering droplets. It shares Geyser Flat with six of its fitful companions.

131

BORA BORA

Emerald isle in an amethyst ocean

Shady palm trees line Bora Bora's broad sandy beaches while a trio of rugged mountain peaks looks down upon the tranquil blue waters of a lagoon. A coral reef fringes the island like a necklace and protects it from the largest of Pacific waves. Such are the qualities that led American author James Michener to describe Bora Bora as 'the most beautiful island in the world'.

Lying 225km (140mi) northwest of Tahiti, Bora Bora is a tiny paradise, 6.5km (4mi) by 4km (2.5mi). Yet it is merely one of a multitude of volcanic islands scattered across the South Pacific. Formed after a series of volcanic eruptions on the ocean floor caused lava to break through the ocean surface, it once towered some 1,200m (4,000ft) above the surrounding sea and 5,400m (18,000ft) above its base.

When the volcano became extinct, no fresh lava added to Bora Bora's size and the forces of erosion took over. Wind and rain lashed the peak, wearing softer rocks away and leaving harder stone to form a tall, jagged outline. The sea invaded the crater, pounding away at its shoreline. The peaks of Taimanu and Pahia, rising more than 655m (2,100ft) above sea level, are all that remain of the north rim of the crater. Riddled with sheer cliffs and steep chasms, the mountain slopes are today clothed in rich, tropical vegetation. Yet the volcano is subsiding as well as being eroded. In a few hundred thousand years its remnants will disappear beneath the water of the lagoon.

As erosion weathered and shaped the contours of the extinct volcano, new forces were at work beneath the coastal waters. Floating in the tropical seas of the Pacific are a multitude of tiny coral larvae. Relatives of the sea anemone, these animals drift in ocean currents until they find purchase on a solid shelf where the waters are clear, shallow and well-oxygenated.

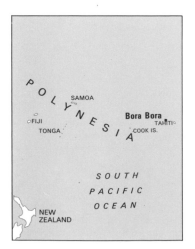

Bora Bora is one of the 14 Society Islands which are part of French Polynesia in the central South Pacific (*left*). At the heart of the island, and covered in tropical forest, stand the remnants of a volcano (*right*). A turquoise lagoon rings the volcano, its waters kept still by the almost circular coral reef which keeps the Pacific breakers at bay. The volcano will eventually subside beneath the lagoon, leaving an atoll – a ring of coral – as the only testimony to its existence.

Each coral larva metamorphoses into a cup-shaped polyp about 2mm (0.08in) across. The upper side of the polyp is open and serves as a mouth; a ring of tentacles catches microscopic food particles and directs them into the opening. The polyps extract calcium from seawater in order to develop hard limestone shells around the lower halves of their bodies. Once established in a favourable spot, the polyps rapidly divide to make new individuals and form a spreading mass of stony coral.

Over thousands of years, these simple animals have formed an almost circular coral reef around the underwater slopes of Bora Bora's volcano. Appearing as a continuous chain of small islands called *motus*, this barrier reef protects the lagoon's tranquil waters from large Pacific waves. The only gap, Teavanui Pass, provides a passage for ships into the island's inner sanctum.

The reef is rich and colourful. Branching coral polyps form extensive but delicate forests while bulky types form great masses of solid limestone. Myriad animals depend on them, not only as a place to hide from predators or to wait for prey, but also as a primary source of food. Shellfish, such as sea snails, cowries and bivalves, move slowly through the coral gardens while giant clams, up to 1.2m (4ft) across, filter the water for food. Sharks and barracudas, often more than 2m (6.6ft) long, lurk menacingly in the shallows in search of prey.

Colonizers of paradise

Long-distance Polynesian seafarers were the first to set foot on Bora Bora some time before the 9th century AD. Their ancestors were the Lapita, an early race of skilled navigators who reached Fiji from New Guinea around 1000 BC. No one yet knows for certain why the Polynesians, or even their forefathers, set off on the long journeys for which they have become renowned.

Descriptions of Polynesian society given by early explorers suggest voyages between nearby islands were common. Polynesians regularly put to sea in their hardy canoes either to fish or to travel. Canoes caught in storms were frequently blown far from home, forcing occupants to seek refuge on unknown islands. In addition, Polynesian society was often split by inter-tribal warfare, leaving defeated tribes no option but to take to their canoes and seek new territory.

Once at sea, Polynesians are adept at finding land. They can interpret wave patterns, ocean swells and cloud formations, reading the signs that tell them if land is near. The presence of an island interferes with the normal wave pattern: by judging the rocking motion of his canoe, a skilled pilot can deduce the distance and direction of the island.

The Polynesians cleared a small part of the tropical forest on Bora Bora, farmed the land and fished the waters of the reef. During the 18th century, European ships made temporary stops at the island during their long voyages of exploration. In 1769, British explorer Captain James Cook made the first record of the island – he named it Bolabola – though he never landed there.

Bora Bora may not remain an island idyll for long. During World War II, more than 5,000 US troops were stationed there, bringing with them electricity, money and a new way of life. The islanders abandoned their tradition of growing copra and vanilla, but were at a loss when the Americans left in 1946. The 1970s witnessed an increase in population and employment, but also the arrival of movie moguls and property developers. One developer who had built a condominium complex on Bora Bora's north shore hinted at the island's fate when he said: 'There's a travel boom in the South Pacific that's just beginning, and it's going to go all the way from Pitcairn to Papua.'

A palm-capped reef of coral and sand, measuring some 14.5km (9mi) by 9.5km (6mi), rings the still waters of Bora Bora's lagoon. In the background, and silhouetted against the sky, rises the island's mountainous backbone, composed of red-black volcanic rock. In traditional outrigger canoes, a small navy of islanders fishes the clear blue waters of the lagoon. Modern canoes, though unchanged in style, are smaller than those of times past. Polynesian craft of the 18th century were recorded as being double-hulled, propelled by long paddles and capable of holding 40 people each.

A small Catholic church in Vaitape, Bora Bora's chief town, symbolizes the conversion of the island's pagan Polynesians to Christianity. Since 1847, the island has been governed by the French after they annexed the 14 Society Islands, of which Bora Bora is a part. The French also annexed a further four archipelagos in Polynesia – the Australs, the Marquesas, the Tuatomus and the Gambiers – and brought over the majority of their populations to the Catholic Church.

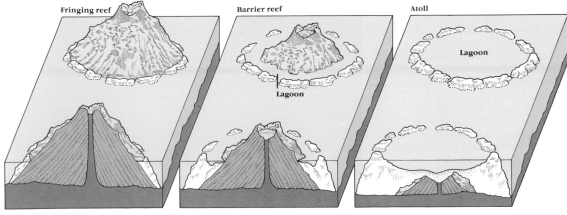

Three types of coral reef are involved in the sequential development of a tropical lagoon. First, fringing reefs evolve on the shores of a volcanic island. As the land subsides, the coral grows upward into a barrier reef separated from the central island by a lagoon. Bora Bora is presently at this second stage. When, eventually, the island subsides beneath the sea, the circular reef encloses the lagoon and becomes known as an atoll.

135

SAN ANDREAS FAULT

The time bomb that threatens to split California apart

At first sight, the streets of Taft in central California are like those in any other American town. Broad avenues are lined by houses with neat gardens; automobiles are parked along the sides of the roads and the street lamps stand at regular intervals. But something is out of place: the lamps are not quite in line, and occasionally a whole road is strangely kinked as if one end were slowly sliding past the other.

These distortions arise because California's bedrock is not at rest. Its Pacific coast is moving northwest along the San Andreas Fault, while much of the state's landscape is shifting southeast. This relative movement does not seem particularly fast – the rate is around 5cm (2in) a year. But it means that Los Angeles will reach San Francisco's latitude in around 10 million years time.

Every Californian knows about the San Andreas Fault: it comes with the territory. From San Francisco to where the Colorado River meets the Mexican border – a distance of almost 1,125km (700mi) – Californians are living above a nest of earthquakes, many imperceptible except to the 'antennae' of scientific instruments. A few, however, such as the 1906 San Francisco earthquake, are devastating.

Two tectonic plates, or sections of the Earth's crust, meet along the San Andreas Fault. Much of North America rests on the American Plate, while most of California's coast lies on the Pacific Plate. Each of the world's 12 major plates is effectively a gigantic raft some 96km (60mi) thick and as old as the Earth itself – around 4,600 million years. As the viscous fluid of the Earth's inner mantle continually circulates, the plates jostle for position.

When they collide head-on the plates throw up huge mountain ranges such as the Alps and the Himalayas. Plates that move apart flood the land with water, creating areas such as the Red Sea and

The San Andreas Fault slices the territory of California in two, from San Francisco to the point where the Colorado River crosses the Mexican border (*left*). Like a scar that cannot heal, the fault line slashes across the landscape. On the Carrizo Plain (*right*), 160km (100mi) north of Los Angeles, streams attempting to hold their course are offset by successive earthquakes. In one instance a stream bed has been offset by as much as 130m (396ft).

Lake Baikal in the USSR. At the San Andreas Fault, the plates slide inexorably past each other and breed earthquakes.

Around 250 million years ago, the Earth's envelope of tectonic plates was configured in such a way that the continents of the world formed one supercontinent, which has become known as Pangaea. When, 50 million years later, Pangaea split up, the continents drifted apart. Geologists estimate that, at around this time, the American and Pacific plates began to move.

The earthquake zones

The San Andreas Fault runs in a smooth arc across California. From the south it skirts the northern edge of the Salton Sea, a lake 72m (236ft) below sea level, before climbing to the crest of the San Bernardino Mountains. Heading north, the fault line runs to the west of the San Jaoquin Valley, crosses the Coast Ranges and descends to the Pacific Ocean at San Francisco.

For much of the fault's length, the two adjacent plates grind past each other at an average rate of 3.5cm (1.5in) a year. The energy generated in these 'creeping' zones of the fault is released in many thousands of tiny earthquakes, which can be detected by sensitive instruments but do little damage. But in the 'lock' zones of the fault, the two plates seem to stick together and mutually prevent each other from moving. Huge geological stresses accumulate over a period of 100 to 200 years before the cohesion is overcome and the plates unlock. When this happens vast amounts of energy are released and earthquakes measuring as high as 7 on the Richter Scale result. The 'quakes of 1857 (at Fort Tejon, north of Los Angeles) and 1906 (at San Francisco) were examples.

Between the creeping and the lock zones lie 'intermediate' zones in which there is minor, but regular, earthquake activity. Parkfield, a small town midway between Los Angeles and San Francisco, has an approximately 22-year cycle of earthquakes measuring magnitude 6 on the Richter Scale. The earthquake of 1857 and subsequent ones of 1881, 1901, 1922, 1934 and 1966, all 'broke' the same part of the fault, a stretch around 25–30km (16–19mi) long that descends 10km (6mi) into the Earth. Nowhere else in the world is the same earthquake repeated time and again. The next 'quake in this zone is expected before 1993.

Predicting the unpredictable

From evidence in the rocks around the San Andreas Fault, geologists estimate that there have been 12 large earthquakes in the region since AD 200. Californians know there will be more in the years to come, but they have little idea of precise timings. Nor do they know exactly where or how large they will be – the worst scenario predicted by the civil engineers who design buildings in Los Angeles is a 'quake of magnitude 7 on the Richter Scale. Such an earthquake in southern California would kill, it is estimated, between 17 and 20 thousand people, cause some $69 billion damage to property, and create major fire and toxic problems for 11.5 million of its inhabitants.

Few of the world's geological features are studied more intensively than the San Andreas Fault. Instruments measuring water pressure in wells 75m (246ft) deep can distinguish between solid Earth tides and slippage in the fault. Two-colour lasers measure the relative movement between one side of the fault and the other. Information from the array of measuring devices focused on the land's every move is fed via satellite to computers programmed to respond to abnormal signals. The hope is that the state of California can be alerted to earthquakes, such as the one anticipated at Parkfield, up to 72 hours before they happen.

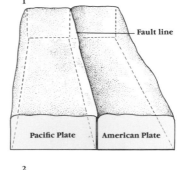

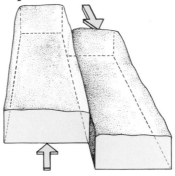

A fault line exists where two sections of the Earth's rocky crust, or lithosphere, adjoin (**1**). Friction between these tectonic plates temporarily delays their natural movement in opposite directions, but accumulating stress energy deforms material around the fault below the surface.

Along the San Andreas Fault, the American Plate is sliding southeast, while the Pacific Plate is moving northwest (**2**). In an earthquake, the stress energy is violently released, and the plates move. Afterward, displacement can be seen on the surface where the plates have slipped past one another before resettling.

A monstrous, rucked-up seam traversing the otherwise featureless Carrizo Plain displays the San Andreas Fault at its most obvious. For much of its length, from Mexico to northwest California, the fault is relatively inconspicuous. This section occurs about 40km (25mi) north of Fort Tejon, a town which in 1857 was the epicentre of one of the worst earthquakes in historic times; the earth was ruptured for 320km (200mi).

Grotesque contortions in the upthrust earth (*right*) reveal the awesome physical stresses in the fault line area. This phenomenon, found on Highway 14 at Palmdale, California, inland north of Los Angeles, is called a parallel auxiliary fault warp. Although most of the displacement caused along faults is horizontal, the deformation of these layers of sediment, with their weirdly folded appearance, is an example of vertical movement of the Earth's crust.

Edwardian figures in a devastated cityscape stroll down Sacramento Street, surveying the aftermath of the great San Francisco 'quake of 1906. The land movement caused a horizontal displacement of 6.4m (21ft) on either side of the fault.

In the late 1980s, nearly five million people occupy the San Francisco Bay area, which is riven with three major fault lines and thus targeted by seismologists as a likely setting for the next devastating earthquake.

GRAND CANYON

A corridor through geological time

In 1858, political disputes between the US Government and Mormon settlers intent on moving south from Utah prompted the exploration of the lower Colorado River. The War Department assigned army officer Lieutenant Joseph Ives to lead the expedition into the territory of northwest Arizona. In a stern-wheeled steamer named *Explorer*, Ives took two months to travel up the river from its mouth on the Gulf of California before his party had to abandon the boat and venture overland.

At the south rim of what he called 'the Big Cañon of the Colorado', Ives rode a mule along a ledge that was 'within three inches of the brink of a sheer gulf a thousand feet deep; on the other side, nearly touching my knee, was an almost vertical wall rising to an enormous altitude'. In general, Ives was unimpressed by the grandeur of the sights he explored. In his report, he wrote: 'Ours has been the first, and will doubtless be the last party of whites to visit this profitless locality. It seems intended by nature that the River Colorado, along the greater part of its lonely and majestic way, shall be forever unvisited and undisturbed.'

Ives and his men were not the first white men to witness the majesty of the Grand Canyon – and they certainly were not the last. In 1540, a young Spanish caballero, Francisco de Coronado, led a band of 300 armed men into Grand Canyon territory in search of gold and other bounty. On hearing of a great river to the west, Coronado despatched one of his captains, Garcia de Cardenas, to investigate.

Coronado's historian wrote that the patrol spent three days on the 'bank' – known today as the South Rim of the canyon – and tried to find a 'passage down to the river, which looked from above as if the water was six feet across, although the Indians said it was a half a league wide . . .'

The Grand Canyon covers around 5,200sq.km (2,000sq.mi) of territory – roughly the size of the state of Delaware – along the Colorado River in northwest Arizona (*left*). One of the most isolated parts of America, the canyon lies around 480km (300mi) east of Las Vegas and 112km (70mi) north of Flagstaff. Eroded from a flat plateau by the torrential river, the canyon (*right*) reveals layers of rock that date back 2,000 million years to the Precambrian Era.

To anyone standing on the Grand Canyon's South Rim the view is like the edge of the world. At its deepest, a point known as Granite Gorge, the canyon drops away for 1.6km (1mi); at its widest it spans 29km (18mi). At first sight it is impossible to appreciate the immensity of the yawning chasm; rank upon rank of precipitous walls and rugged spines of rock split the landscape into a complex labyrinth of fantasy canyons.

The canyon's formation

The Grand Canyon rates as the world's largest gorge, and extends for 444km (277mi) along the Colorado River. Its walls have been created almost entirely by the action of the river's waters. The banded sides of the immense chasm are like a corridor through geological time: as layer upon layer of rock was exposed by the river's power of erosion, so the canyon grew deeper and the earliest history of the Earth was laid bare.

Of the four eras of Earth's history, only the first two – the Precambrian and the Palaeozoic – are represented in the Grand Canyon's walls. The rocky layers of the two most recent eras (the Mesozoic and Kainozoic) have been worn away. The schist rocks at the bottom of Granite Gorge are 2,000 million years old and were formed at a time when the molten core of the Earth punctured its surface.

The uppermost layers of the canyon date from the Palaeozoic Era which began around 600 million years ago. Composed of various sandstones, shales and limestones, these layers were laid down as sediment at the bottom of an ocean. The most recent layer is 250 million years old. Around 10 million years ago, the ocean dried up and was replaced by a flat plain across which wound a broad river, known today as the Colorado. Upheavals in the Earth's crust caused the rocks of northern Arizona and southern Utah to push upward in a large dome.

Around two million years ago, the Colorado narrowed, its waters thus increasing in power. With ever greater force the river battered the ground, wearing away the base of its valley as quickly as the surrounding rocks bulged upward. Consequently, the course of the river remained roughly the same while the walls of rock through which it flowed grew higher and higher.

The expeditions of a one-armed explorer

In 1869, the area of the middle reaches of the Colorado was shown on maps as an empty space of roughly 259,000sq.km (100,000sq.mi). To remedy this omission, a geologist from Illinois, Major John Wesley Powell, set out to explore these uncharted regions. Despite losing an arm in the Civil War, Wesley had become a skilled explorer in the Rocky Mountains.

On May 24, 1869, nine men and four wooden boats left Green River Station in Wyoming and journeyed down the upper Colorado. They took three months to reach the Grand Canyon, carefully mapping the river and the surrounding land. But in the process they lost one boat, together with its important cargo of scientific instruments, and had only one month's food supply left. For several days they were swept onward by the Colorado's swift currents and turbulent white water.

When they halted at a broad beach deep in the heart of the canyon, three men abandoned the expedition. But, after scaling the walls to the canyon's rim, they were cut down by a band of hostile Indians. On August 29, the remaining party of six hungry men and two battered boats emerged into previously charted waters at the western end of the canyon. Two years later, a better-equipped expedition enabled Powell to chart the Grand Canyon and put the world's most rugged river profile on the map.

Colossal totem poles hewn by nature from the eroded rock make this a popular lookout point on the canyon's top edge. Here, guard rails are provided, but for most of the gigantic gorge's vertiginous perimeter, a sheer drop awaits the unwary.

Descent of the canyon by mule is along a switchback trail cutting through successive layers of geological history – some two million years' worth.

The few paths that have been maintained and developed were initially picked out by the bighorn sheep and native deer whose instincts for a sensible route down were subsequently exploited by prospectors and Indians. The modern trails, no narrower than 1.5m (5ft) are generous by comparison to the width of the originals.

Upstream northeast of Grand Canyon the Colorado River flows between the lofty granite walls of Glen Canyon. The Glen Canyon Dam to the east interferes with the river's annual flooding and subsidence, by which means it had formerly rebuilt its eroded shores. Before the dam's completion, the river brought a daily average of 500,000 tonnes of rocks, pebbles and sand into Grand Canyon. When the river was in full flood, this enormous quantity was multiplied by more than 50.

METEOR CRATER

Gigantic footprint of a cosmic missile

Viewed from the flat, barren plain that surrounds it, Meteor Crater is little more than a low unremarkable hill. But whoever ambles up its gentle gradient greets the unexpected. For at the crest of the hill there yawns a steep-sided, saucer-shaped bowl of immense proportions: 1,265m (4,150ft) in diameter and, at its maximum, 175m (575ft) deep.

News of the gaping hole in the desert near Arizona's Canyon Diablo was first reported in 1871. It was generally regarded as a volcanic crater since there were others, such as the still-active Sunset Crater, in the same region. In the 1890s, America's leading geologists began investigating the phenomenon after discoveries by mineralogists of iron fragments, even diamonds, suggested a meteorite was involved in the crater's creation.

The geologists, notably G.K. Gilbert (1843–1918), concluded that the crater had been caused by 'a steam explosion of volcanic origin'. Nevertheless, the idea that a gigantic meteor had collided with the Earth attracted many adherents. Foremost among them was Daniel Barringer (1860–1929), a Philadelphia mining engineer who explored the site in 1903. Convinced the meteorite was buried beneath the crater, he purchased the land and, in 1906, began drilling. Although the central core eluded him, Barringer and his team discovered enough iron and nickel-iron fragments to persuade the scientific world that the crater was probably formed by a meteor.

The size of the extra-terrestrial missile that struck the Arizona desert remains a subject of speculation. In the 1930s, scientists estimated its weight at 14 million tonnes and put its diameter at 122m (400ft). Later calculations reduced these figures to only about 2 million tonnes and a diameter of 79m (260ft). By current estimates the meteor was even smaller still, with a weight of

Meteor Crater lies 30km (19mi) to the west of Winslow on the plains around Arizona's Canyon Diablo (*left*). The almost circular rim of the crater, created by the impact of an enormous meteor, rises more than 45m (150ft) above the surrounding plain (*right*). Sediments discovered at depths of more than 30m (100ft) have led geologists to conclude that, around 12,000 years ago, the crater had been a rain-filled lake. The drier climate of more recent times has evaporated the lake and turned the region into an arid lunar landscape.

70,000 tonnes and a diameter of 25m (80ft).

But the collision was cataclysmic: to create such an immense crater, the meteor must have been travelling at a velocity of around 48,000kph (29,960mph), generating a blast equivalent to half a million tonnes of TNT. By comparison, the atomic bomb that destroyed Hiroshima in 1945 possessed a force equivalent to 20,000 tonnes of TNT. The meteor's fall-out must have devastated an area with a radius of more than 160km (100mi) from the point of impact. The collision probably occurred around 22,000 years ago, although some people think the crater was formed around the time of Christ while others put its age at 50,000 years.

Evidence of the meteorite

'The most interesting spot in the world' declared the distinguished Swedish chemist Svente Arrhenius (1859–1927) of the crater after he had visited it at the turn of the 20th century. He was seeking, but did not find, evidence to support his theory that life is spread throughout the universe via microscopic spores carried on meteorites.

During the 1930s, around $400,000 was spent on drilling bores into the floor of the crater. Fragments of nickel-iron believed to have come from the meteorite were found at depths of 260m (700ft). Below this the rock was undisturbed. All attempts at finding the core intact below the crater have been abandoned. Scientists now believe the meteor exploded on impact, and that much of its material vaporized into the air.

The millions of nickel-iron grains discovered at the site are thought to have condensed from a hot metallic cloud that resulted from the blast. In addition, individual nickel-iron fragments as heavy as 640kg (1,400lbs) have been found scattered over an area of 260sq.km (100sq.mi). Any doubts about the crater's origin were removed in 1960 with the discovery there of coesite and stishovite, two rare forms of silica that can only be created under high temperatures and pressures – conditions generated by a meteor colliding with the sandstone desert.

Meteorites around the world

In the past, people who witnessed meteors falling through the sky, or who had found evidence of them on the ground, were traditionally treated with scepticism. Former US President Thomas Jefferson (1743–1826) remarked in 1801 that he would 'sooner believe that two Yankee professors had lied than that stones had fallen from the sky'. In the following two years scientists discovered the similarity in chemical composition of fragments from all over the world, establishing once and for all the existence of meteors.

Deep space is, in fact, full of meteors orbiting the sun in groups. Many are smaller than coins while others weigh millions of tonnes. Hundreds of thousands shower the Earth's atmosphere daily but are often so small they either go unnoticed or else are burned up and seen as shooting stars. The best known meteor showers are the Leonids: these arrive in November and appear to come from the constellation of Leo. The largest meteors retain enough of their central core to strike the Earth with tremendous force. On impact they become known as meteorites: thus Meteor Crater should properly be called Meteorite Crater.

Chubb Lake in Northern Ungava, Canada, is the largest meteorite crater in the world. Blasted from solid granite, it is almost 3.2km (2mi) across and filled with a lake 244m (800ft) deep. The largest known meteorite found on the surface of the Earth lies near Grootfontein in Namibia. Composed of solid iron, it measures 2.7m (9ft) by 2.4m (8ft) and weighs 60 tonnes.

Access to the floor of Meteor Crater is gained via a steep trail that may take an hour to descend. The sandstone floor is rich in minerals formed by the high temperatures and pressures created by the meteor's impact.

Until 1967, when Meteor Crater was declared a National Natural Landmark, the only commercial exploitation of the minerals was the mining of its silica, which is considered to be among the purest in the world.

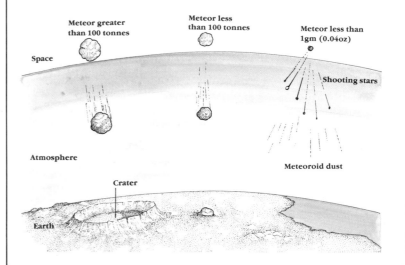

Deep space is full of meteors, or interplanetary particles, which regularly enter the Earth's atmosphere. Meteors weighing more than 100 tonnes occasionally penetrate the barrier formed by the atmosphere: they explode on impact and form a crater. Meteors weighing less than 100 tonnes but more than 1gm (0.04oz) may strike the Earth without exploding. Meteors weighing less than 1gm (0.04oz) are burned up and seen as shooting stars, before falling to Earth as meteoroid dust.

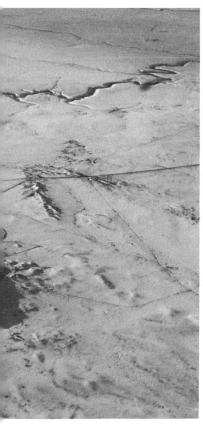

The impact crater in the Arizona desert is one of around 30 such structures on the Earth's surface. They are far more common on the Moon, which does not have an atmosphere to protect it from the constant barrage of meteors. Data collected by various spacecraft indicate that the Moon is riddled with an astonishing three million million craters.

The enormous meteorite (*right*), known as Peko 2, was discovered in the mid-1980s in China's eastern province of Shandong. The Chinese authorities estimated the meteorite fell around 1,400 years ago, and claimed that its weight of four tonnes made it the largest meteorite ever found on Earth. However, the meteorite discovered in 1920 near Grootfontein in Namibia was 15 times heavier.

RAINBOW BRIDGE

Nature's monumental sandstone arch

A Paiute Indian named Nasjah Begay was responsible for leading the first white men to Nonnezoshi, 'the hole in the rock'. Each member of the three-man party had been intrigued by rumours about a great stone arch in the vicinity of Navajo Mountain, a peak shaped like the back of a whale. A highly visible landmark from all directions, this mountain was situated in some of the most inaccessible country in North America.

On August 14, 1909, Dr Byron Cummings of the University of Utah, government surveyor W.B. Douglass, and John Wetherill, proprietor of the Oljeto Trading Post at the northern end of Monument Valley, set eyes on Rainbow Bridge. According to various reports, Cummings was the first to see it, Douglass the first to reach it and Wetherill the first to walk under it. US President Theodore Roosevelt (1858–1919) would later exclaim that it was the greatest natural wonder in the world. The US President of the time, William Taft, preserved it for posterity when, on May 30, 1910, he declared 65 hectares (160 acres) of land around Rainbow Bridge a national monument.

The pink sandstone of Rainbow Bridge curves gracefully above a canyon to a height of 94m (309ft), the highest natural arch in the world and enough to encompass the Capitol in Washington D.C. At the peak of its arc, the arch is 13m (42ft) deep and 10m (33ft) wide, sufficient width to run a two-lane highway across. The span across the canyon measures 85m (278ft), almost as long as an American football field.

Local Paiute and Navajo Indians had long conceived of the arch as a 'rainbow turned to stone', not only because its shape resembled the most perfect arc of all, the rainbow, but also because of the delicate hues of its constituent sandstone. Under a bright desert sky the stone is tinted dark lavender, while in the

Rainbow Bridge is located on the Navajo Indian Reservation in southern Utah, beside the state border with Arizona and some 240km (150mi) north of Flagstaff (*left*). Resembling the arc of a rainbow, the pink sandstone arch (*right*) spans the meandering Bridge Creek. This stream flows via the Forbidding Canyon into the Colorado River from its source at Navajo Mountain, some 8km (5mi) to the southeast of Rainbow Bridge.

late afternoon sun, the sandstone arch is splashed with a variety of reds and browns.

Around 80 million years ago, many streams started to wind their way across a sandstone plateau from the northern slopes of Navajo Mountain, a peak 3,166m (10,388ft) high. When, some 15 million years later, the entire region was lifted up into a gradual dome by upheavals in the Earth's crust, the streams carved deeper and deeper courses for themselves. In the stream known today as Bridge Creek, a large but isolated buttress of rock jutted out from the side of the canyon, forcing the waters to flow around it.

Alternate extremes of daytime heat and night-time cold flaked the sides of the sandstone rock, thinning it to a narrow window. Rock debris carried by the stream battered the base of the buttress and weakened it. Eventually, the irresistible force of the stream's water found a way through, widening the window and exposing its arch to the polishing effects of wind and weather. Yet the stream continued to carve the canyon below it, making the arch seem higher and higher.

The 1909 party that discovered Rainbow Bridge was forced to travel on horseback across some of the most inhospitable territory in the United States. It was because of this extreme inaccessibility that the fabulous arch took so long to be discovered. However, in 1963, engineers completed the Glen Canyon Dam, creating Lake Powell in the process and pushing water from the Colorado River into the 90-odd adjacent canyons. When Lake Powell, which has a 3,200km (2,000mi) coastline of red rock, reaches full capacity a narrow tongue of water pushes its way down Bridge Creek and under the giant arch. As a result, visitors to Rainbow Bridge can journey by boat to within shouting distance of the monument.

Utah's land of arches

Rainbow Bridge is only one of hundreds of sandstone arches in Utah. Some 300km (180mi) to the northeast, near the Mormon farming town of Moab, lies Arches National Park. Here, in a surreal landscape, sit 83 arches of various shapes and sizes. In contrast to Rainbow Bridge, the windows of these arches have been honed out of the rock by rainwater, wind and sand.

The Entrada sandstone rock from which the arches were formed was laid down around 40 million years ago. Softer yet flakier than the common Navajo sandstone of southern Utah and northern Arizona, this rock collects rainwater in its cracks and fissures. As the water freezes and melts, the cracks widen, so loosening great fragments of rock; flash floods add muscle to the enlargement process. As this weathering continues, windows appear in ridges and outcrops; high winds and driving sand complete the finishing touches to the smoothly polished arches.

Landscape Arch has a span of 89m (291ft), making it 4m (13ft) longer than Rainbow Bridge and therefore the longest natural arch in the world. As fragile as a collar-bone, the sandstone fin of Landscape Arch projects from rugged outcrops of rock at an average height of 30m (100ft) above the floor of a canyon. In geological terms, the arch will not last long since one stretch of its rock is only 1.8m (6ft) thick and barely able to withstand the merciless power of erosion.

Another of the park's fantastic structures is Delicate Arch, a salmon-pink semicircle of sandstone that resembles the stapes bone of the middle ear. Local farmers and cowboys have nicknamed it 'old maid's bloomers'. From its isolated position on the rim of a desolate rock bowl, the arch neatly frames the snow-capped La Sal Mountains, some 32km (20mi) to the southeast.

Rainbow Bridge began as a solid sandstone buttress projecting into the canyon carved by the stream now called Bridge Creek. The base of the buttress was eroded by silt-laden water (**1**) and by weathering, until an opening was made in the vulnerable stone.

Water then flowed through the short-cut, gradually – along with other erosive forces – enlarging the aperture (**2**). The creek later carved a small canyon below the arch which serves to accentuate the height and emphasize the exceptional symmetry of Rainbow Bridge.

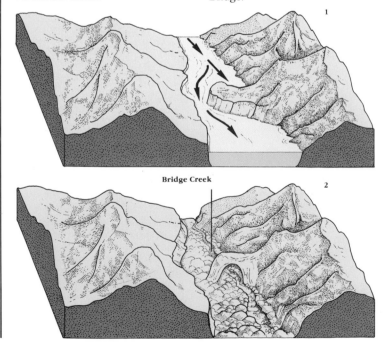

Bridge Creek

150

Landscape Arch, in Utah's
Arches National Park, is the
world's longest natural arch. It
was created not by river action,
but by weathering and sand
erosion of the region's original
91m (300ft)-layer of Entrada
sandstone. This sits atop a darker
red variety of sandstone called the
Carmel formation, which is in
turn underlaid with the harder
Navajo sandstone.

Delicate Arch, also found in
Arches National Park, lives up to
its name in both appearance and
structure. Perhaps the most
remarkable feature of this fragile
sandstone hoop, which stands
26m (85ft) high and has a span of
20m (65ft), is that one of its 'legs'
is only a frail 1.8m (6ft) thick.
This, combined with the relatively
soft rock of which the arch is
formed, means that geologically
its days are numbered.

MONUMENT VALLEY

Rocky red backdrop of the Wild West

Aficionados of American movie director John Ford instantly recognize Monument Valley as the setting for some of his well-known epics: *Stagecoach* (1939), *My Darling Clementine* (1946) and *She Wore a Yellow Ribbon* (1949). In all more than 25 movies have been filmed on this barren, sagebrush plain, where massive outcrops of red sandstone stand in silence like fragmented architectural relics.

More than 200 million years ago, the area now occupied by Monument Valley on the Utah-Arizona border was a windswept desert of red sand. When the land was inundated by a sea, heavy deposits of mud fell as sediment to the sea bed and compressed the sand into red sandstone. The thick layers of mud became shale. When, around 65 million years ago, the area was uplifted the sea bed became a huge, flat plateau of sandstone covered by a thin layer of shale and conglomerate, a kind of hard, sedimentary gravel.

The forces of erosion started to work at once. Where earth movements had opened up cracks and fissures in the shale, water and wind etched their way in to the underlying sandstone and bore it away. Clefts in the rock were deepened and broadened until they formed a labyrinth of canyons and gullies. As a result, sandstone terraces were whittled down into rocky tablelands or mesas which, in turn, were reduced to monuments.

Geologists use the term 'monuments' to describe remnants of erosion which, because they are higher than they are broad, often resemble manmade buildings or artefacts – pillars, spires, chimneys, skyscrapers, castles, temples. Hence the name of the region, which American western writer Zane Grey (1875–1939) described as a 'yellow-and-purple corrugated world of distance'.

Many monuments reach heights approaching 300m (1,000ft),

Monument Valley straddles the Utah-Arizona state border, some 43km (27mi) north of Kayenta and about twice as far southwest of Bluff (*left*). The valley is scattered with giant sandstone buttes and massifs; many have conical bases which indicate erosion is still at work. Some monuments, when silhouetted against the sky, resemble familiar images, such as the two Mittens (*right*). The Merrick Butte (to the right of the photograph) commemorates the death of a silver prospector murdered by Indians in the 1880s.

and are composed of a vertical shaft of red sandstone topped by a cap of more resistant rock which has protected the lower layers from erosion. Often the monuments are surrounded by a spreading cone of sandstone debris that has fallen from the existing structure and is an indication that erosion is still in progress. These final remnants of a huge sandstone plateau will themselves be flattened by erosion within a few thousand years.

The shapes of the various monuments have inspired a potpourri of names. The impressive massif of Castle Rock is crowned by crenellations which might be mistaken for battlements. Hen-in-a-nest bears a fanciful similarity to a squatting fowl. Perhaps the most aptly named are The Mittens: two gigantic formations standing close together, each with a narrow column of rock, the 'thumb', beside a broad butte of rock, the 'fingers'.

In the land of the Navajo

A thousand years ago, the lands straddling the Arizona-Utah border were inhabited by Indian tribes living in pueblos, villages of mud-brick houses of some sophistication. These Indians, known as the Anasazi or 'ancient ones', laboured hard with the help of complex irrigation systems to cultivate the arid desert. But when the climate became even more arid in the middle of the 12th century, the Anasazi were forced to abandon their homes and migrate southward. Their ruined dwellings can still be seen in Monument Valley. One dwelling, known as the House of Many Hands, bears hundreds of palm prints made by hands dipped in white paint.

In the 16th century, the lands formerly populated by the Anasazi Indians were inhabited by the Hopi, Zuni and the Navajo tribes. The semi-arid plains around Monument Valley were occupied largely by the Navajo, the most populous of all American Indian groups. They developed a pastoral life of herding goats and sheep. Much of their culture is inherited from the Anasazi – their turquoise jewellery and the geometrical designs on their blankets, as well as their sand paintings which are used to help cure diseases.

The Navajo Indians had a reputation for raiding their neighbours, whether these were other Indian tribes, Mexicans or white settlers. When, in 1859, a few hot-blooded Navajo braves raided a white settlement, the might of the US Cavalry was brought to bear upon them. Many Navajo took refuge in Monument Valley and the surrounding territory. But, in 1868, after a protracted war during which frontiersman Kit Carson fought alongside the Cavalry, the Navajo surrendered and were transported to Fort Sumner in New Mexico. In 1874, Chief Manuelito led a delegation to Washington and persuaded the US Government to return most of the confiscated lands, including Monument Valley, to the Navajo Indians.

During World War II, valuable sources of weapons-grade vanadium were discovered and mined from the cap rocks on top of some of the sandstone buttes, such as Elephants' Feet near Tonalea at the southwestern entrance to Monument Valley. Since the war, deposits of uranium have also been found in the area, which is still owned and administered by the Navajo, adding considerably to the Indians' revenue.

In the 1980s, the Navajo continue to herd sheep and goats in Monument Valley, which they think of as 'the place among the red rocks'. The Navajo also work silver and turquoise mined from their land, and weave blankets with their traditional geometrical designs from the animals' wool. It is said that every Navajo weaver places a deliberate mistake in each blanket, since the Indians believe that perfection marks the end of a weaver's life.

Rocky steeples shimmering in the desert heat are all that remain of a huge rocky tableland of sandstone. The Totem Pole (the pinnacle at the right of the picture) stands 165m (500ft) tall, almost as high as the Washington Monument. At sunset, it casts a shadow almost 56km (35mi) long across the valley floor. The cluster of spires is known as Yei Bichei.

The solitary pinnacle of Chimney Rock resembles a finger pointing heavenward. Like the adjacent group of monuments, the spire is capped by hard rock, beneath which is a thin shaft of soft sandstone. The rocky debris resulting from millenia of weathering lies in a conical heap at the pinnacle's base, a reminder of the fate which will soon befall Chimney Rock itself.

The saguaro cactus (*Cereus giganteus*), symbol of the Wild West, grows up to 15m (50ft) high and may live for more than 150 years. Shaped like a candelabra, the cactus is peculiar to southwestern USA and northwest Mexico. Its white blossom, which opens at night, is the state flower of Arizona; its crimson fruit provides Indians with a vital source of food.

PETRIFIED FOREST

Stone trees from the age of the dinosaurs

Dawn breaks over the Painted Desert in northeastern Arizona. In the trembling violet light, the contours of hills, cones and gullies begin to separate from the darkness. As sunlight adds a greater clarity, tinted bands can be distinguished on the sandstone hills, as if their different strata has been meticulously painted in pale, glinting hues.

Within this strange fairy world, the trunks of conifers lie scattered over the parched ground like huge corks, most of them broken up into great broad logs. But if anyone tried to incise their name into one of these trees, they would get a shock. For their bark, sap and woody flesh have been turned to solid stone. Moreover, colour sparkles from these ancient fragments: hexagonal quartz crystals have transformed their woody molds into jewelled caskets of orange, blue, yellow and pink.

The Petrified Forest, first reported by army officer Lt. Lorenzo Sitgreaves in 1851, is the largest assemblage of fossilized trees in the world. Another petrified forest, at Sigri on the Greek island of Lesbos, has nothing like the size, amount or otherworldly setting of the stone trees of Arizona. To the Navajo Indians, the scattered trunks were the bones of a legendary giant, Yietso; while the Paiutes believed them to be arrow shafts from the quiver of their thunder god, Shinauv.

The history of these iron-hard logs began around 200 million years ago when the desert was not a sculpture of hot and gleaming rock, but a broad, swampy floodplain. Instead of today's bobcat, coyote and badger, dinosaurs roamed among the giant conifers that populated the slopes of hills and volcanic mountains. Ninety per cent of the trees rose to 30m (100ft) with a diameter of around 2m (6.6ft). Some towered to twice this size. Relatives of the more modern Norfolk Island pine (*Araucaria*

The Petrified Forest covers an area of over 388sq.km (150sq.mi) and lies 72km (45mi) east of Winslow, in the southwestern corner of Arizona's Painted Desert (*left*). Scattered around the tinted, wrinkled hills of Blue Mesa (*right*), the once great conifers are now an assortment of broken trunks, logs and chips of petrified wood. Infused with silica and other minerals, many remnants of the forest have been transformed into a glittering kaleidoscope of crystalline quartz.

heterophylla) and shaped like mushrooms, these trees are given the scientific name of *Araucarioxylon arizonicum*. The remainder of the trees rarely grew taller than 21m (70ft). They belonged to two distinct genera: *Woodworthia*, which resembled modern pines, and the aquatic *Schilderia*, which had buttresses like those of the swamp cypress.

Metamorphosis of trees to stone

In the course of time, the trees died and their huge, wrinkled torsos were swept away periodically by floodwater. Shepherded into rushing, swollen rivers they would become trapped in gullies or locked nose to tail in massive logjams. They were covered with thick deposits of mud and sand, and ash from the fallout of nearby volcanoes.

Deprived of oxygen in their marshy envelope, the trees began their strange metamorphosis. Groundwater absorbed silica from the volcanic ash as well as other minerals from the sediments. This mineral-rich water soaked into the tissues of the incarcerated tree trunks, permeating cell walls with molecules of silica. These either turned into quartz or, when mixed with other minerals, crystallized into semiprecious quartz jewels, such as agate, jasper, onyx, carnelian or amethyst.

As the trees became stone, the minerals took on the shape of the wood cells and multiplied into precise stone copies of the trees. Or else they replaced the wood cells and formed huge, inexact replicas of the trees. As millions of years passed, new sediments covered the logs' coating of mud, ash and silt, turning it into hard layers of shale and sandstone that geologists call the Chinle Formation. By this time, the logs were entombed up to a depth of 300m (1,000ft).

Exhumation of the forest

Around 65 million years ago, the process began that would expose the once giant conifers in a glittering new guise. A slow but massive upheaval of the Earth's crust caused the Rocky Mountains to rise, pushing up the ancient burial ground of the mineralized trees at least 1.6km (1mi). Exposed to the elements, the newer, less resistant sediments were eroded away. Eventually, the layers of shale and sandstone around the petrified logs were also removed, leaving the ancient conifers to take their place again under the hot Arizona sun.

Dinosaur fossils were exposed and are still found on the desert floor, evidence of the creatures that roamed the land when the conifers were alive. Kings of the water were the phytosaurs, huge crocodile-like reptiles with slim, elongated snouts. They preyed on metoposaurs, swamp-dwelling amphibians up to 3m (10ft) long with enormous heads and stocky, powerful legs. These fish eaters weighed more than half a tonne and, like the modern hippopotamus, revelled in shallow water.

Around the same size were the aetosaurs, heavily-armoured animals resembling armadillos but with reptilian features. They preyed on *Placerias*, a kind of three-eyed rhinoceros with large tusks. The fossil teeth of this creature reveal a vegetarian diet. Weighing more than two tonnes and measuring up to 3.5m (11.5ft) in length, the slow-moving, gregarious *Placerias* dug up roots and plants with a pair of large tusks.

The exhumation of the trees and other fossils continues as erosion wears away the soil. Although the desert receives only 22.5cm (9in) of rain a year, most of it comes as brief but violent thunderstorms that etch away up to 2.5cm (1in) of soil annually. Like the Badlands of South Dakota, the rain washes away the soft sandstone and turns the sunbaked clay of the desert into a temporary quagmire.

(1) The mighty conifers which became the Petrified Forest at first studded a vast flood plain, active with volcanoes, some 200 million years ago.

(2) When the trees died, from a variety of natural causes, some were carried off by flood waters, forming logjams at the bends in rivers.

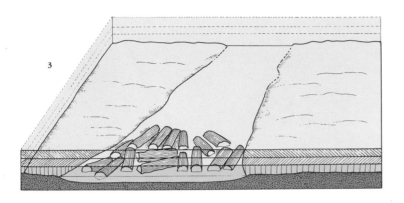

(3) The logjams were quickly covered by thick layers of mud, sand and volcanic ash. Silica in solution from the ash seeped into the wood and crystallized.

(4) Further sediments piled up on the buried logs, creating layers of shale and sandstone, but erosions and upheaval later exposed the petrified wood.

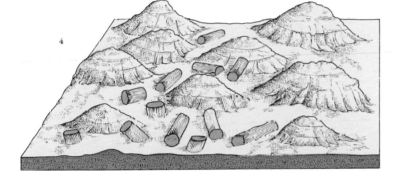

A petrified log seen in cross-section perfectly preserves all the key characteristics of the original wood. The annual rings, converted to quartz, tell the ancient tree's life story, a botanical biography millions of years old.

Resembling gem-hard fire logs, segments of petrified wood lie strewn across a strange and lovely desertscape. What time and erosion chopped into chunks will next break down into chips, and ultimately into grains of quartz. Nature's toll was increased by souvenir hunters who took away thousands of tonnes of the stone before the Petrified Forest was protected by law.

WHITE SANDS

A porcelain desert glazed with dunes

In the noonday heat of New Mexico, the White Sands shimmer like a mirage. Soft and cool to the touch, these enchanting dunes wedged between dull brown stoneware mountains, are not composed of silica, the chief constituent of 'ordinary' sand, but of gypsum. Covering an area of 712sq.km (275sq.mi), the White Sands are the world's largest surface deposit of this mineral.

Gypsum – known chemically as calcium sulphate – is one of the Earth's more common minerals and has served man for thousands of years. The ancient Egyptians plastered the vaults of their pyramids with it. The ancient Greeks made windows out of transparent gypsum crystals known as selenite. In the 20th century, the mineral is used to make plaster, wallboard and, as plaster of Paris, moulds and casts for broken limbs.

Gypsum is produced commercially only from large deposits that are close to urban areas, such as those in Texas and Utah – which explains why no one has staked a claim at the White Sands. Instead, this unique habitat was preserved by the Federal Government when former US President Herbert Hoover declared, in 1933, that 570sq.km (220sq.mi) of the gypsum desert should become a national monument.

Around 100 million years ago the shallow seas covering much of the southwest USA started to retreat, leaving behind lakes of saltwater which gradually evaporated in the sun. As well as ordinary salt, gypsum precipitated from the mineral-rich solution and formed thick layers on the old sea bed. When, around 65 million years ago, the San Andres and Sacramento Mountains were created (at the same time as the Rocky Mountains) by upheavals in the Earth's crust, further beds of gypsum were exposed to the power of the elements.

Rain and meltwater from the mountains dissolved the gypsum

The White Sands lie in the Tularosa Basin in New Mexico's southwest, between the San Andres Mountains to the west and the Sacramento Mountains to the east (*left*). Of the few plants hardy enough to withstand the rigours of the climate and the shifting gypsum dunes, the yucca (*right*) is the most prominent. This unusual member of the lily family has tough leaves that restrict water loss and roots that develop new growth when the plant is moved on by the dunes.

and washed a concentrated solution of the mineral to a lake whose deep basin had no outlet. This lake, known as Lake Lucero, lies at the foot of the San Andres Mountains and is the source of the White Sands. Water draining into the lake is trapped and is evaporated by the combined effects of warm temperatures and steady winds. As a result, a crust of glassy selenite crystals is formed on the lake's surface.

Prevailing southwesterly winds convert the delicate sheets of selenite into tiny gypsum grains and deposit them farther up the Tularosa Basin to the northeast of the lake. The loose grains, which turn to powder when rubbed between the fingers, are piled up to form dunes often as high as 15m (50ft) above the desert floor. But the winds do not leave the gypsum at rest. Instead, they shift the dunes around the basin at a maximum of 10m (33ft) per year.

Wildlife of the dunes

Although the White Sands are constantly on the move, some plants do manage to gain a foothold – a survey in the 1950s accounted for more than 100 different species. These plants, such as the yucca, sumac and cottonwood, are able to survive the unstable, alkaline sands and the almost continuous dryness. Because of the movement of the dunes, the roots of some of these plants, especially those of the cottonwood, may reach over 30m (100ft) long.

Prominent amongst these hardy plants is the candle of the Lord (*Yucca elata*), so named because Spanish settlers likened its spike of waxy, bell-like flowers to flaming tapers. The fibre of its tough leaves was once used by Indians to make rope and baskets. The same Indians ate the shoots, ground the seeds and made soap from the roots. Modern cattle ranchers often substitute the yucca for forage when droughts make grass scarce.

The yucca has a remarkable one-to-one relationship with a specific moth: the yucca can only be pollinated by the yucca moth (*Tegiticula yuccasella*) and the moth will lay its eggs nowhere else but in the yucca's flowers. When the flowers open, the white moth collects the pollen and rolls it into a ball with its legs. The moth then seeks out another bloom: it places the pollen ball on the stigma, thus fertilizing the flower, and lays between one and four eggs at the base of the bloom.

The moth's eggs and the yucca's seed develop together. When the larva hatches it eats about half of the ripened seeds. After a few days, it munches its way out of the yucca, drops to the ground and burrows into the loose gypsum where it pupates. After a year, the adult moth emerges.

Permanent animal residents of the White Sands are few. The bleached earless lizard (*Holbrookia maculata ruthveni*), and the white Apache pocket mouse (*Perognathus apache*), which is nocturnal and rarely seen, are found nowhere else in the world. Horned lizards (*Phrynosoma spp.*) have the ability to change their colour to match their surroundings. On the usual brown deserts of New Mexico, they are patterned with shades of brown. On the basalt lava flows which dot the region, only black horned lizards are found. On the glaring gypsum surface of the White Sands, the lizards are pure white.

At the margins of the White Sands plant and animal life is more abundant. The gold fetid flowers of the buffalo gourd bloom beside pink centauriums and purple sand verbenas. Coyotes, skunks, gophers, badgers, kangaroo rats, snakes and porcupines all make their home here. Occasionally these creatures venture into the dunes at night, their tracks clearly visible in the morning sun like footprints in newly fallen snow.

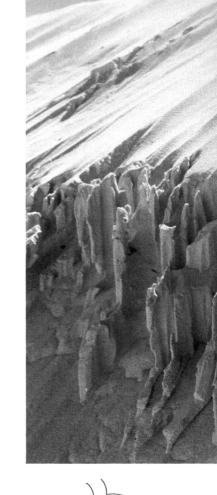

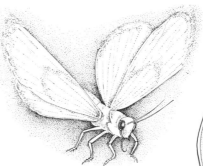

The yucca moth (*Tegiticula yuccasella*) has evolved a special relationship with the yucca plant (*Yucca elata*). The white moth (*above left*) needs yucca seeds as a food source for its larvae; the plant needs the insect to pollinate its flowers. This relationship of mutual benefit is known as symbiosis.

A female moth, equipped with specialized tentacles under her head, gathers pollen from one of the yucca's long stamens (*above*

right) and rolls it into a ball. After laying an egg in the ovary of the flower, the moth deposits the pollen ball on the stigma, thus ensuring the flower's pollination and the development of the seed.

The soft, white crystals of gypsum are easily crushed to a powder when rubbed between finger and thumb. The prevailing southwesterly winds at White Sands also crush the crystals into grains. As they sweep across the desert they regularly fashion fantastic sculptures from the windward sides of gypsum dunes.

The bleached earless lizard (*Holbrookia maculata ruthveni*) is unique to the dunes of White Sands. Normally, the earless lizards of New Mexico's Tularosa Basin are brown, but in the gypsum desert natural selection has favoured the palest members of the group and only these survive. Curious, friendly and able to run fast, the earless lizard will shed its tail if seized and then grow a new one.

BADLANDS

Barren vistas of a landscape in decay

The Badlands of South Dakota could be the typical setting for a science fiction movie. Steep ridges are riven by deep ravines, their tinted, horizontal bands scored by dozens of vertical channels. Narrow, flat-topped hills tower high into the air, in places capped by a jumble of rocks and boulders. A plateau wizened by erosion, the landscape is unlike any other in the world, except perhaps Arizona's Petrified Forest.

The local Sioux Indians called this landscape *mako sica*, meaning 'bad land'. In the 18th century, French fur trappers from Canada translated this as *les mauvaises terres à traverser*, meaning 'bad lands to cross', because after the rains the slippery surface of clay made travel treacherous. Later, English-speaking ranchers coined the now familiar name of Badlands.

Originally, the Badlands were the lands beside the White River in South Dakota, an eighth of which, the Badlands National Monument, achieved national park status in 1978. However, a similar landscape in North Dakota has also acquired the name. Now the Theodore Roosevelt National Park, these Badlands occupy 285sq.km (110sq.mi) beside the Little Missouri River. Former US President Theodore Roosevelt (1858–1919) was referring to them when he wrote: 'There are few sensations I prefer to that of galloping over these rolling limitless prairies, rifle in hand, or winding my way among the barren, fantastic and grimly picturesque deserts of the so-called Bad Lands.'

Around 80 million years ago, the region occupied by the Badlands of South Dakota was a shallow sea, its bed full of rich and varied sediments. The same upheaval in the Earth's crust that formed the Rocky Mountains around 65 million years ago crumpled the sea bed and pushed it upward. At first an immense, marshy plain, the area became transformed into a rolling prairie

The Badlands of South Dakota occupy an area of 15,550sq.km (6,000sq.mi) – roughly the size of the state of Hawaii – between Mount Rushmore and the Missouri River (*left*). Torrential spring rains continue the erosion started many thousands of years ago. Layers of sedimentary rock that once made up a flat plain are washed away into the White River, leaving behind a treacherous and otherworldly landscape (*right*).

covered in rich green grass and patches of coniferous forest.

Fossils trapped in the sedimentary rocks reveal the rich fauna that inhabited the region in its succession from sea to fertile plains. Marine turtle shells are one of the most common fossils found in the Badlands – the largest was 3.7m (12ft) long. Equally prevalent are fossils of oreodonts, piglike mammals with large teeth designed for ruminating. The largest animal was probably *Brontotherium*. This lumbering herbivore stood over 2.5m (8ft) tall at the shoulder and walked on four pillar-like legs. On the snout of its tiny head was a long, thin horn of solid bone bisected at the tip into two branches. Less than half the size was *Mesohippus*, a small, three-toed ancestor of the modern horse, its thin body and long legs clearly built for speed.

Erosion by torrential rains

Wind, freezing temperatures and running water are the potent forces that have transformed a rich pastureland into a decaying landscape. Devoid of vegetation, the rocks of many hills crumble at the slightest touch. Nearly every footfall loosens stone fragments, which tumble down into the ravines. Pinnacles of rock, such as Vampire Peak, are being reduced in height by as much as 15cm (6in) every year.

The semi-arid Badlands are simultaneously being shaped and fast eroded by the torrential rains of spring and early summer. Only 38cm (15in) of rain falls each year, but most of it arrives in intermittent downpours of tremendous violence. The Badlands become a network of gushing streams and powerful rivers. Water cascades from ridges and rushes through ravines with astonishing force. Because the ancient sedimentary layers have never been compressed by harder rock, their soft, tinted substance is easily invaded and washed away by running water. The runoff joins the White River and then the Missouri, taking with it a huge cargo of clay, stones and gravel.

The White River gains its name from the Badlands' chalky sediments which do not settle but remain permanently in solution. The chief source of chalk is the Wall, a desolate stretch of cliffs and buttresses around 60m (200ft) high. The only creatures surviving in this treacherous place are reptiles, such as rattlesnakes and bull snakes, bats, and rodents, such as the Badlands chipmunk.

Elsewhere in this wild landscape, living things find it equally hard to gain a permanent foothold. A site that has provided a niche for a generation may disappear overnight. Yet sheltered pockets do exist, especially on the fringes of the Badlands where the soil is more stable. Here plants, such as the buffalo grass and prairie golden pea, and animals, such as the prairie dog, can thrive.

Prairie dogs (*Cynomys ludovicianus*) are burrowing, squirrel-like rodents that inhabit underground colonies or 'towns'. Composed of a highly ordered network of tunnels and chambers, these towns have designated areas for sleeping, defaecating and food storage. One of the prairie dog's predators is the rare black-footed ferret (*Putorius nigripes*), a distant relative of the true ferret unknown to science before 1851.

The Badlands' story is not only one of decay and struggle for survival. In 1963, 53 bison were reintroduced; by the early 1980s, the herd contained more than 300 animals. In 1964, bighorn sheep were brought in. When agriculture was banned from the Badlands in 1978, a variety of prairie grasses began to take hold in the uncertain soil; and following in their wake came the almost-extinct pronghorn (*Antilocapra americana*), an antelope-like creature renowned for being America's fastest mammal.

The ancient layers of sediment of the Badlands, softer than true stone, are easy for nature to carve. Water, wind and frost are the tools that continuously gouge out this forlorn and ravaged terrain, starved of vegetation except in isolated pockets. The subtly contrasting bands exposed down the sides of the eroded hills attest to millennia of varied deposits, including silt from river overflow and ash from volcanic eruptions.

The pronghorn (*Antilocapra americana*) is recovering from near extinction in outlying Badlands' areas where suitable prairie grasses have now reestablished themselves. The antelopes all but fly across their open habitat, able to cover up to 8m (27ft) in a single, extended stride. Top speed in short bursts is about 86kph (55mph); 70kph (45mph) can be maintained for roughly 6.4km (4mi).

Badlands National Park was originally intended to protect the wealth of extraordinary rock formations and fossils the area contains. But no human intervention can alleviate the progressive natural disintegration of these fragile clay spires.

OKEFENOKEE SWAMP

The trembling ground of Georgia's wildlife sanctuary

The Okefenokee is a territory of deception. What appears to be firm ground is, in fact, afloat. Clumps of tall trees look as though they grow from solid ground yet have their roots below water. Broad swathes of long grass, which bend like wheat to gusts of wind, turn out to be reeds and rushes. Wide meadows of grass and wild flowers are little more than bobbing masses of vegetation. Yet despite the apparent fragility of its situation the Okefenokee's unique wetland habitat supports a remarkable collection of wildlife.

When the Atlantic Ocean receded from southeastern Georgia and Florida, it left a lake of salt water in a shallow, saucer-shaped depression. Layers of clay and silt accumulated on the limestone bed, while on top of these decaying vegetation turned to peat. Today the water resembles tea, stained by the tannic acid from cypress trees and peat. But it is not stagnant. Water from several natural springs moves slowly toward and along two rivers: the St Mary's leading to the Atlantic and the Suwannee flowing into the Gulf of Mexico.

Although seemingly bottomless, the brown water of the Okefenokee is rarely more than 1m (3.3ft) deep. Its shallowness means that many plants, such as water lilies and floating hearts, can root in the mud and push their stems to the air and sunlight. Swamp cypress trees (*Taxodium distichum*) are the most spectacular. Often festooned with Spanish moss, they grow in large clusters throughout the swamp and are highly adapted for survival in soggy and unstable soil.

The base of the cypress trunk flares out to form many buttresses, which effectively spread the weight of the tree over a larger area of mud. Its roots, unable to gather oxygen from the waterlogged soil, send up nodules or 'knees' through the water

The Okefenokee Swamp occupies 1,760sq.km (680sq.mi), mostly in southeastern Georgia but also extending into the northern tip of Florida (*left*). In the tea-coloured waters of the swamp, groves of tall cypress trees anchor floating islands to the peat bed beneath (*right*). In spring and summer, large water lilies and clusters of floating hearts transform the surface of the water with a multitude of white and yellow blooms. Many species of heron and egret feed on swamp fishes.

to the air. While it may take a thousand years for a cypress to reach old age, a mature specimen will eventually grow too large for the mud that supports it and is likely to topple over during high winds. But the cypress will not die as long as its root system stays intact. Branches will grow up from the trunk, prolonging the life of the fallen giant by many years.

Throughout the Okefenokee, strange islands float in the forests and in the open marshes or 'prairies' where the maiden cane grass grows in profusion. Composed either of peat or of tree branches, matted moss and grass tangled together to form rafts, these islands give the swamp its name. Anyone walking across them finds that they shift and undulate alarmingly. The local Seminole Indians called them *owaquaphenoga*, meaning 'trembling ground'. In the 19th century, white settlers rendered the word as okefenokee.

The protected wetland refuge

The animal community thriving in the swamp's aquatic environment has an immense variety – from snakes, lizards and turtles to a multitude of frogs and toads; from ospreys, herons and the threatened sandhill cranes to a surfeit of woodpeckers, warblers and red-winged blackbirds. The king of the swamp is the alligator (*Alligator mississippiensis*). Its young are only 10cm (4in) long when they hatch yet they hunt at once, voraciously feeding on any frog, fish or insect they can tackle. Full-grown alligators may reach 4m (13ft) in length and weigh 225kg (496lbs).

The all-pervading water creates ideal conditions for amphibians, which need to lay their eggs in an aquatic environment. Twenty-two species of frogs and toads thrive in the swamp. The southern leopard frog (*Rana pipiens*) inhabits rich low-lying vegetation and is much preyed upon. The squirrel tree frog (*Hyla squirella*) is elusive by comparison – only 2.5cm (1in) long, it can change colour to suit its environment. Moreover, its flattened toes are specially adapted for climbing cypress trees.

The yellow-bellied sapsucker (*Sphyrapicus varius*) is probably the noisiest inhabitant of the swamp. This woodpecker drills rows of holes in the trunks of trees, returning every once in a while to drink the sap and feed on the insects gathered there. Three species of egret flourish again in Okefenokee's refuge, having once been threatened by the great demand for their plumage in early 20th-century hats. Six species of heron are commonly sighted stalking in the shallow water and darting their long necks outward to capture fish in their narrow beaks. The grebes are such poor fliers that, when danger threatens, they often prefer to dive into the water than take to the air.

The fate of the swamp's entire wildlife was jeopardized once the Seminole Indians were evicted by the US Army in 1838 and exiled to Florida. Many settlers tried to turn the swamp into agricultural land, with some success. In the 1890s, Captain Harry Jackson of Atlanta attempted to drain the swamp and create the Suwannee Canal that would link the Atlantic Ocean with the Gulf of Mexico. The project was aborted halfway through and has become known as 'Jackson's folly'.

A lumber company decimated the cypress population of the swamp between 1908 and 1926, all but destroying the wildlife and the livelihood of the settlers. When the Federal Government bought the ravaged land in the 1930s they converted it into a wildlife refuge. The Okefenokee has since restored and healed itself. Today, tourists are entranced by the wonders of its flora and fauna, fishermen come in search of bass, bluegill and chain pickerel while campers take to canoes and enjoy the serenity of a primeval wilderness.

The huge American alligator (*Alligator mississippiensis*) dominates the swamp in both size and power, preying upon a wide range of the inhabitants. Its reputation for wiliness is well founded. Alligators, and their close relatives the crocodiles have the most advanced brains of any living reptiles, despite their primitive appearance.

A navigable swathe through the lilypad prairie gives no hint of the Okefenokee's depth. The water, stained a peaty brown, conceals the bottom of what is, in fact, a relatively shallow depression. While the surface may appear still, there are barely perceptible currents draining this unique watershed in opposite directions toward outlets 240km (150mi) apart. The swamp is too far from the sea to be subtropical; in winter, it is not immune to the rigours of frost.

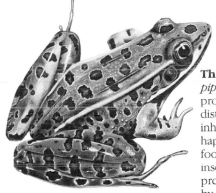

The leopard frog (*Rana pipiens*), with a slim body, prominent back ridges and distinctive spots, is a common inhabitant of the swamp. It will happily feed on nearly any animal food Okefenokee affords – insects and crustaceans for preference but is in turn hunted by many swamp residents.

The great egret (*Casmerodius albus*) is a long-legged, keen-eyed bird, but unlike its relatives among the heron and egret families, it does not remain motionless to hunt. Rather, it stalks slowly along the water's edge, seeking fish to spear with its formidable black bill.

NIAGARA FALLS

A mighty cascade doomed to perish

Foolhardy souls no longer plunge over Niagara Falls in barrels. The huge boulders that broke away from the lip of Niagara's precipice before World War II rendered such barrel shooting suicidal. And the rock collapse also provided an insight into how the falls were created and how they will be destroyed. Each time the precipice collapses, the position of the falls moves upstream. Ultimately, they will cease to exist.

The falls probably reached their present position almost 700 years ago when Goat Island separated them into two distinct cascades. To the east of this wooded isle, on the United States' side, are the American Falls: 56m (184ft) tall and 323m (1,060ft) across, these falls carry less than 10 per cent of the Niagara River's water. The remainder hurtles over the Horseshoe Falls on the Canadian side: 54m (176ft) tall, these falls are 675m (2,215ft) in length and, not surprisingly, are horseshoe-shaped. Because the river is deeper on the Canadian side, erosion is faster. As a result, the Horseshoe Falls have retreated an estimated 300m (1,000ft) in 300 years.

Until around 10,000 years ago, the enormous glaciers covering much of North America forced the waters of the Great Lakes to flow south to the Mississippi River. When the glaciers melted at the end of the ice age, the Niagara River was formed. It drained the waters of Lakes Superior, Huron and Michigan, via Lake Erie, across a small plateau and over a steep escarpment into the basin of Lake Ontario.

The Niagara plateau is composed of several horizontal layers of different rocks. At the bottom is a thick bed of shale overlain by sandstone. Both kinds of rock are comparatively soft and therefore easily eroded. Above them lies dolomite, an extremely hard and weather-resistant type of limestone. The turbulence

The Niagara Falls lie on the Canada–US border between Lake Erie and Lake Ontario (*left*). The waters of Lake Erie, which collect currents from Lakes Huron, Superior and Michigan, flow along the Niagara River at an average rate of around 5.7 million litres (1.25 million gallons) a second. When it reaches the Niagara Falls, the river splits into the American Falls and the larger Horseshoe Falls (*right*), generating a spectacular display of rainbows in the pure white mist and spray.

created at the foot of the falls quickly erodes the shale and sandstone, leaving a dolomite overhang.

The result of this erosion is a vertical drop over which the Niagara River plunges. When the weight of unsupported dolomite is great enough, the overhang collapses and falls down as huge boulders to the bottom of the escarpment. This constant undercutting began when the Niagara River was formed. As a result, the mighty falls are now around 11km (7mi) upstream from their original position.

Lake Ontario stands 98m (320ft) below the level of Lake Erie. Linking them together, the Niagara River carries, at full flow, around 514 million litres (113 million gallons) each minute. But since engineers can now control this mighty volume, only around half flows over the falls at any one moment; the remainder is channelled into the manmade Welland River and into hydroelectric power schemes. Because this diversion of water considerably reduces erosion, geologists estimate the falls will take much longer than Nature intended to retreat the 48km (30mi) across the entire Niagara plateau. In short, Lake Erie and Lake Ontario will merge and the Niagara Falls will disappear in about 25,000 years time.

As the falls are eroded upstream, a deep gorge about 90m (295ft) wide is left behind. The river surges between its towering walls at around 40kph (25mph). Some 4.8km (3mi) downstream from the falls the river makes an abrupt right hand turn, creating a whirlpool through which no boat can safely pass. On July 24, 1883, Captain Matthew Webb, the first man to swim the English Channel, attempted to swim the length of the gorge but was dragged under the whirlpool and drowned.

Niagara's charms

Around 100 of the world's waterfalls are higher than Niagara, and at least two carry a greater flow of water. But few are more enchanting. Because the waters of the Niagara River are free of sediment and sand, they create a permanent cloud of mist and pure white spray as they cascade over Niagara's precipice. To quote one 19th-century traveller, the water is 'green as emerald and clear as crystal'.

After visiting Niagara's spectacle in 1842, the English novelist Charles Dickens (1812–1870) wrote: 'What voices spoke from out those thundering waters; what faces, faded from Earth, looked out upon me from its gleaming depths; what heavenly promise glistened in those angels' tears, the drops of many hues that showered around and twined themselves about the gorgeous arches which the changing rainbows make.'

The renown of Niagara Falls as a place for barrel shooters and other stuntmen lured the famous French acrobat Charles Blondin (1824–1897) in 1859. Hoping to outdo all previous daredevils, Blondin walked a tightrope strung across the falls in four minutes — and in front of thousands of spectators. A few days later, he made the trip blindfolded, then returned to carry his manager on his back. Finally, the following year, he walked across on stilts.

Niagara Falls has not only attracted writers and daredevils; its spectacular natural beauty has also drawn romantics and honeymooners. The first recorded couple to visit Niagara were Jerome Bonaparte, younger brother of the French Emperor Napoleon, and his bride Elizabeth Patterson in 1803. Many couples have followed in their footsteps, establishing a custom which prompted Irish playwright Oscar Wilde (1854–1900) to utter caustically: 'Niagara Falls is the second great disappointment of an American bride'.

A bird's-eye view of the Niagara plateau shows the course of the Niagara River from Lake Erie (*below*). The original position of the falls was on the escarpment, some 11km (7mi) downstream. The plateau's inclination toward Lake Erie tends to hasten the overall erosion of the falls.

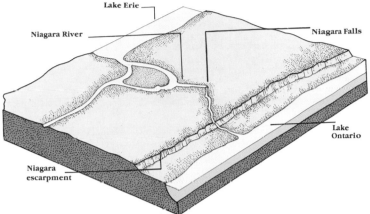

The position of Niagara Falls creeps away from Lake Ontario upstream toward Lake Erie. Over the last 300 years, the Canadian Falls have changed from a slightly concave to a wholly horseshoe shape (*below*), in the last one hundred years, their rate of retreat has averaged almost 1.2m (4ft) per year.

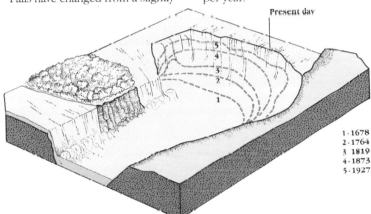

1 - 1678
2 - 1764
3 - 1819
4 - 1873
5 - 1927

The permanent mist shrouding Niagara Falls has become one of its attractions, especially on clear days when rainbows shimmer in the air. A tale relates how local Indians annually hurled their fairest maiden over the falls to appease the thunder god living below. The story is commemorated by boats named 'Maid in the Mist', which in the summer venture into the blizzards of spray at the brink of Niagara's turbulence.

An enormous weir built halfway across the Niagara River on the Canadian side prevents water from flowing over the Horseshoe Falls (*above*) and diverts it toward the American Falls. The weir forms part of a system that channels water into one of the largest hydroelectric complexes in the western hemisphere.

Under terms agreed in a 1950 treaty, Canada and the United States share this runoff equally. At night and during the winter, around three quarters of Niagara's water is diverted for electricity generation. During daylight hours in the summer, the diverted water equals half the river's full volume.

The turbulent impact of Niagara's water (*below*) erodes the falls as it strikes the foot of the precipice. The constant battering eats away the weak shales and sandstones, leaving the dolomite rock unsupported at the top. Eventually, this tough layer of limestone fractures and crashes down.

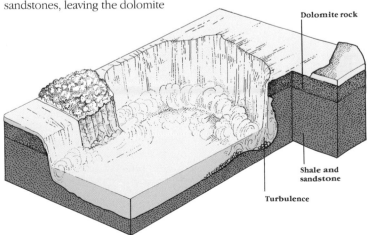

Dolomite rock

Shale and sandstone

Turbulence

RIVER AMAZON

The mightiest waterway on Earth

In 1542, a band of 60 Spaniards led by Captain Francisco de Orellana became the first Europeans to journey down the Amazon from the heart of the Peruvian jungle. Having sailed down the Napo River from Ecuador in search of food, they found themselves drawn into waters that filled them with awe. A friar, Gaspar de Carvajal, recorded their impressions: 'It came on with such fury and with so great an onrush that it was enough to fill one with the greatest fear . . . and it was so wide from bank to bank from here on that it seemed as though we were navigating launched upon a vast sea.'

Following in the footsteps of the Spaniards came the Portuguese, one of whom, Pedro Teixeira, took 2,000 men and 47 boats from the river's mouth to Quito, Ecuador, in 1638. European scientists braved the so-called Green Hell to bring back details of the mighty river. French mathematician and geodesist Charles Marie de la Condamine mapped the Amazon in 1743 and measured its depth, gradient and speed. In 1800, German naturalist Alexander von Humboldt mapped the stream linking the Amazon with the Orinoco River.

British naturalists of the Victorian era were particularly drawn by the rich variety of plants and animals. The discoveries Alfred Russel Wallace made in the Amazon jungles in the early 1850s led him to the same theory of evolution by natural selection as his fellow scientist Charles Darwin. Between 1848 and 1859 Henry Bates collected almost 15,000 animal specimens, half of which were entirely new species. And, at the same time, botanist Richard Spruce collected 7,000 new plant species.

The source of the world's greatest river is Lake Lauricocha, which lies at an altitude of 5,180m (17,000ft) in the Peruvian Andes. From here to the Atlantic Ocean, and under the guise of

The River Amazon rises from Lake Lauricocha in the snowcapped Andes Mountains of Peru, only 192km (120mi) from the Pacific Ocean (*left*). Together with its 1,100 tributaries, the Amazon drains a flat-bottomed basin which covers 6.5 million sq.km (2.5 million sq.mi), almost as large as the continent of Australia. The pale waters of one of its tributaries, a branch of the River Ucayali, meanders through the primary rain forest in eastern Peru (*right*).

many names, the Amazon flows a total of 6,440km (4,000mi), making it the world's second longest river after the Nile.

The Amazon, however, is not simply a single river but an integrated system of rivers, jungles and climate. For the 4,800km (3,000mi) from the foothills of the Andes to the Atlantic, the forest-covered Amazon basin never rises more than 198m (650ft) above sea level. The basin is so flat that the river gradient is rarely more then 2.5cm (1in) in every 1.6km (1mi).

The volume of water in this flat-bottomed basin is staggering. More than 1,100 tributaries, 17 of which are longer than Europe's Rhine, feed the main trunk of the Amazon with storm- and melt-water from the surrounding mountains. At any moment the Amazon and all its tributaries contain two thirds of all the Earth's river water. And in a single day, the Amazon disgorges into the ocean what England's River Thames delivers in a year. So great is this volume that it pushes back the Atlantic's saltwater for more than 160km (100mi), creating a huge freshwater lake.

The waters of the Amazon's greatest tributary, the Rio Negro, are as black as its name suggests. This hue is caused by an acid from the rotting vegetation of the Colombian swamps where the river draws much of its water. At Manaus, where it meets the white waters of the River Solimões to form the Amazon proper, the Rio Negro is 18km (11mi) wide. And such is the speed of the two rivers that they do not truly merge for 6km (4mi). From Manaus the Amazon still has a third of its course to run. As it approaches the Atlantic coast it splinters into channels – the main channel is 50km (30mi) broad – and creates a multitude of islands. Marajó, the largest of these, is approximately the size of Switzerland.

The teeming waters

More than 1,500 different species of fish are known to live in the waters of the Amazon. This figure is 10 times that of all Europe's rivers combined and three times that of the fishes in the River Zaire in Africa. One such species, the pirarucu (*Arapaima gigas*), is reputed to be the world's largest freshwater fish. Weighing in, on average, at a massive 200kg (440lbs) and measuring as much as 5m (16.5ft) long, this bony-tongued fish lives in the poorly-oxygenated swamp waters of the river.

The red piranha (*Serrasalmus nattereri*) is the Amazon's most notorious fish. Although they grow only to a maximum length of 30cm (12in), piranhas hunt in groups and, with their razor-sharp teeth, have been known to reduce large, but wounded, animals to a bare skeleton within seconds. This feeding behaviour is, however, exceptional. The staple piranha diet consists of other fish as well as seeds and fruit.

The largest predator in the Amazon basin is the black caiman (*Melanosuchus niger*), a highly endangered species which can grow 4.5m (15ft) long and has been known to attack human beings. Its usual diet consists of aquatic mammals, such as the manatee, and forest-dwellers, such as the capybara and the tapir, which venture down to the water's edge to drink.

This alligator is far from being the only Amazon species threatened with extinction. Whole families of animals and plants are endangered because of the destruction of the Amazon's rain forest. The enormous diversity of animal and plant species in the Amazon jungle makes it the greatest natural resource in the world. In one acre of primary forest there are about 60 different tree species, some 15 times the number found in temperate forests. Yet, in the name of progress, almost 130,000sq.km (50,000sq.mi) of forest, an area the size of New York State, was destroyed in the single decade between 1975 and 1985.

Canoes, which are vital to the survival of many Amazonian Indian tribes, double as a means of transport and as fishing craft.

After mandioca, the starch-bearing cassava plant native to the Amazon, the Indians of the basin eat more fish than any other food.

The Indian population of the Amazon region has declined from an estimated one million in 1500 to less than 100,000 in the late 1980s. At least 150 linguistically distinct tribes remain, many of whom are, like the Krahó Indians (*below*) of northeast Brazil, protected by the National Indian Foundation.

The black waters of the Rio Negro meet the white waters of the River Solimões at Manaus and form the Amazon proper. The two confluent rivers travel so fast that their respective waters do not begin to merge for about 6km (4mi) downstream. Finally, after about 80km (50mi), the white waters predominate.

The blackness of the Rio Negro comes from the rotting vegetation in the Guyanan highlands where the river rises. The colour of the River Solimões is generated by its huge cargo of white silt, which its tributaries wash down from the Andes Mountains. This volume of silt is so large that the Amazon conveys most of it to the Atlantic Ocean where it stains the seawater grey.

GREENLAND

The iceberg factory on top of the globe

When the Icelandic explorer Eric the Red landed on this snow-clad island in AD 982, he gave it the inappropriate name of Greenland as a lure to his compatriots to settle there. Four years later, many responded to his call; thus was established a colony that lasted for 500 years. The Icelanders settled on the southwest coast, where warm currents from the Atlantic Ocean create a milder, rainy climate.

Two thirds of Greenland, the world's largest island, are covered by an ice cap that contains around 10 per cent of all the Earth's ice. If this ice cap were ever to melt, the oceans would rise by as much as 6m (20ft) – with devastating consequences. Its average thickness is 1,500m (5,000ft), although in some places it can be as thick as 5km (3mi). This immense mass of frozen snow results from the climate's severities and the structure of the underlying rock.

Some of the Earth's oldest rocks lie beneath Greenland's ice desert. Samples of the metamorphic rock from Godthåb, the country's capital, have been dated to the early Precambrian Era, around 3,700 million years ago. The same bedrock extends west-ward to form the core of Canada's landmass. Jagged mountain peaks tower more than 2,600m (12,000ft) above a coastline deeply indented by fjords, while inland the ground slopes down to form a vast, saucer-shaped depression.

Snow falls on the ice cap throughout the year. Raging blizzards can sweep its surface, whipping the snow into deep drifts; or steady falls layer innumerable large flakes to a great depth. But however the snow falls, it cannot melt – the temperature is never high enough. As the snow builds up, so it crushes earlier falls beneath it. When the weight of snow creates a great enough pressure, the lower layers become ice. It is this constant accumulation

Greenland lies largely within the Arctic Circle (*left*) and covers an area of 2,175,600sq.km (840,000sq.mi). This is roughly equivalent to twice the size of Norway, Sweden and Finland put together. The total area of coastal land that remains ice-free and habitable is approximately the size of Norway. Most of Greenland's population of around 55,000 lives in towns and villages along the west coast where the climate is relatively mild. A few brave the arctic conditions in towns, such as Angmagssalik (*right*), on the east coast.

of ice that is mainly responsible for the ice cap's huge depth. In the 1950s, scientists from the US Army's Snow, Ice and Permafrost Research Establishment (SIPRE) discovered that ice at a depth of 50m (165ft) represented snow which had fallen at the time of the American Civil War.

The enormous pressures within the apparently solid cap squeeze out the lower layers of ice, pushing them as glaciers through passes in Greenland's coastal ring of mountains. Most flow with imperceptible slowness, but the Jakobhavn Glacier moves toward Disko Bay, 500km (300mi) north of Godthåb, at a rate of 1.2m (4ft) per hour.

When a glacier reaches the sea, it begins to melt. Cracks open up within the ice. As these widen, icebergs break off at the seaward edge and float away with the tides and currents. Throughout each spring and summer, thousands of icebergs drift southward from Greenland's west coast and head for the Grand Banks of Newfoundland and the shipping lanes of the North Atlantic. Two years after a Greenland iceberg sank the *Titanic* on April 15, 1912, the US Coast Guard began to operate the International Ice Patrol. Each day of the iceberg season, ships and aircraft comb the North Atlantic, plotting the positions and courses of the most dangerous bergs, and alerting all shipping in the affected areas.

The Arctic wildlife

Much of Greenland lies within the Arctic Circle. The tundra in the north is snow-buried for most of the year but blossoms with life during the brief summer. For a few weeks each year mosses, lichens and grasses carpet the ground, providing a staple diet for a large range of animals, such as reindeer (*Rangifer tarandus groenlandicus*). In winter, however, reindeer herds must dig through the snow for food and, when necessary, travel over large distances in search of new feeding grounds.

The rare and protected musk ox (*Ovibos moschatus*) leads a similar life to the reindeer. Standing 1.6m (5ft) high, this shaggy, gregarious relative of the sheep is extremely hardy. Throughout the summer weeks, it eats continuously in order to build up its fat reserves which, together with waterproof hair, will enable it to survive the rigours of winter.

Dozens of bird species flock to coastal regions of Greenland to feed on the abundant summer vegetation. The Arctic tern (*Sterna paradisea*) makes the most spectacular journey. During the Greenland winter these birds fly halfway round the world to the Antarctic to take advantage of the southern summer, then return north to breed the following year. By comparison, the grey phalarope (*Phalaropus fulicarius*) flies only as far as Canada in search of warmer winter weather.

Greenland's waters teem with shrimps and plankton, the microscopic plants and animals which anchor most marine food chains. The waters of Disko Bay, for instance, contain some of the largest shrimp beds in the world. The rare bowhead whale (*Balaena mysticetus*), probably the longest of all the whales that filter-feed on plankton, is a frequent visitor to Greenland's shores. Its body grows to lengths of more than 15m (50ft) of which no less than 5m (16.5ft) is made up by the skull.

Huge numbers of fish, such as cod, halibut and salmon, also feed on the plankton. These fish are eaten, in turn, by seals. Both the common seal (*Phoca vitulina*) and the Greenland seal (*Phoca groenlandica*) thrive in the cold sea. They may gather in their thousands when moving to new feeding grounds or during the mating season. Large groups of walruses (*Odobenus rosmarus*) also congregate on Greenland's shores, but unlike the seals, they feed on shellfish and crabs.

Sled dogs still play an important role in Greenland, despite the advent of motorized transport across the white wilderness. Distances between villages are conventionally measured in the time it takes a dogsled to make the journey.

In winter, the fjords used by boats in summer become frozen highways for the sleds. Dog teams snarl at one another as they overtake or pass in opposite directions. It is usual for passengers to hop off the sleds periodically and run alongside – as much to keep warm as to reduce the animals' burden.

At rest, as seen here outside the town of Angmagssalik on the island's bitter cold eastern coast, a sled-party subsides into a tangle of harness and huskies. But a team underway in the hands of a skilful driver is a triumph of graceful coordination.

Jagged mountains rear out of the peaceful waters of Tasermiut Fjord at the island's southern tip, west of Cape Farewell. On a particularly balmy July day, the temperature might climb to 10° C (50° F), but the frigid finger of sea reaching inland never warms up enough to encourage swimming – except by the hardy marine wildlife.

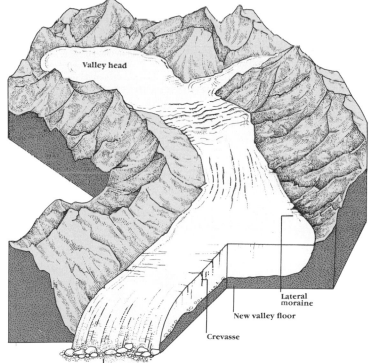

Valley head

Lateral moraine

New valley floor

Crevasse

Terminal moraine

Glaciers form in mountain areas where snow falling faster than it can melt compacts gradually into granular ice. A build-up of pressure sets the frozen mass in slow motion downhill.

Although it may follow the course of a valley originally carved by water, the glacier acts more like a solid than a liquid. Its upper and middle parts move fastest,

while its base and sides are slowed by friction with the valley, which it widens and deepens.

The rocky debris, or moraines, the glacier gathers en route carry out the monumental scouring that transforms the valley's V to a U-shape. Crevasses occur where the ice flows over irregularities. If the glacier pushes as far as the sea, its snout breaks up into icebergs which float away.

LAKE TITICACA

The great lake cradled high in the Andes Mountains

According to legend, Lake Titicaca is the birthplace of the Inca civilization. The sun god instructed his children, Manco Capac and his sister-wife Mama Ocllo, to travel until they found a place where a golden rod would sink into the earth. Having discovered such a place on an island in Lake Titicaca, they gave birth to the Inca race, the 'children of the sun'. This island, known as the Island of the Sun, remains one of the lake's most sacred places and the local Indians still celebrate this 'birthday' with a festival every November 5.

The deep blue waters of Lake Titicaca sparkle on the scenic Altiplano, or 'high plain', between parallel ranges of the snow-capped Andes Mountains. At an altitude of 3,800m (12,500ft), it is the world's highest navigable lake and, at a length of 177km (110mi), it is also one of the longest.

The first steamer to navigate the lake was the 200-tonne *Yaravi*. Built in England in 1862, it was dismantled on the coast of Peru and its components hauled up the Andes on mules and llamas. Reassembled at Puno on the Peruvian side of the lake, it carried passengers and goods for a century. To replace the *Yaravi*, a hydrofoil called the *Inca Arrow* was put into service in 1966. By 1987, there were five such craft plying the waters between Bolivia and Peru.

Much of Lake Titicaca is contained in a steep-sided basin with a maximum depth of 281m (922ft) and an average depth of 107m (351ft). Evidence from the surrounding terrain suggests that, after the last ice age, around 10,000 years ago, the lake's shoreline was 45m (150ft) higher than it is today. In the course of a year, the water level fluctuates between 50cm (1.6ft) and 1m (3.3ft). From December to March, when heavy rains sweep across the region, the level rises; but for the rest of the year the

Lake Titicaca straddles the border between Bolivia and Peru almost 4km (2.5mi) up in the Andes Mountains (*left*). Covering an area of 8,290sq.km (3,200sq.mi) – roughly a third of the size of Lake Erie in the USA – the lake is the largest body of freshwater in South America. Totora reeds growing in the lake's shallow marshy waters are used by the local Indians, as they have been for centuries, to fashion distinctive boats (*right*).

level falls. From year to year, however, the waters of the lake may rise and fall by a total of 5m (16ft).

Just over half the lake's annual water input comes from rainfall. The remainder is brought by many rivers and streams from the surrounding plain – a drainage area more than seven times the size of the lake. Ninety per cent of the lake's water evaporates as a result of the high altitude, strong sun and frequent winds, leaving behind salts and minerals that impart a slightly brackish taste. The only outlet, the Desaguadero River, drains less than 10 per cent of the lake's water – not to the Pacific Ocean, but southward to Lake Poopó which has no outlet. Only evaporation stabilizes its level, and its waters are 20 times as salty as those of Lake Titicaca.

The average temperature of the air around Lake Titicaca is 11°C (51°F) although the daily extremes may fluctuate considerably around this figure. Consequently, the average water temperature of the lake remains fairly constant at around 11°C (51°F). While this is comparatively cold for a lake so near the equator, it is explained by the lake's high altitude. By comparison, Lake Kariba, on the Zambesi River in Africa, has a mean temperature of 22°C (72°F), even though it lies on a similar latitude.

Titicaca's isolation, salinity and constant temperature has not given rise to any highly specialized flora and fauna. Because of the restricted variety of fish, rainbow trout and lake trout were introduced in 1940. Rainbow trout provided good catches for the fishermen for many years, but a parasite they brought with them has since multiplied and caused the demise of several of the native fish species.

Amphibians are the most extraordinary animals of the lake. In 1968, a team led by French underwater explorer Jacques Cousteau investigated the waters of Titicaca. Above a thick layer of mud and slime on the bottom they found a multitude of frogs, some up to 60cm (2ft) long. All were able to breathe through their skins and so consequently they rarely surfaced for air. Cousteau's team concluded that Titicaca's frog population was of the order of one thousand million.

Reed boats of Titicaca

The region's Uru, Quechua and Aymara Indians live a life largely unchanged since the Spanish invasion of the 16th century and continue to use reed boats for travel and fishing. They harvest the totora reeds (*Scirpus tatora*) that grow in profusion among the marshes of the lake, especially in Puno Bay. The canoe-shaped reed boats, often powered by sails made of reed or canvas, are characteristic of Titicaca – yet they bear a striking similarity to those of ancient Egypt.

The resemblance was enough to convince the Norwegian explorer Thor Heyerdahl that Egyptians had brought civilization to South America. Employing two totora craftsmen from Suriqui, one of Titicaca's 36 islands, he constructed a reed craft in Morocco according to Egyptian design. In this boat, which he named *Ra II*, Heyerdahl successfully crossed the Atlantic to Barbados in 1969 and in so doing demonstrated that Egyptians, who would have had the same materials at their disposal, could feasibly have made the same voyage.

For centuries, the Uru Indians have lived near Puno on islands built from totora reeds. Anchored in the shallow waters of Titicaca, these floating 'Uros' islands are equipped with churches and schools as well as houses – the Uru even spread soil on them and plant crops. By adding fresh layers of matted reed to the surface to replace the underside that rots in the water, each island can be made to last for many years.

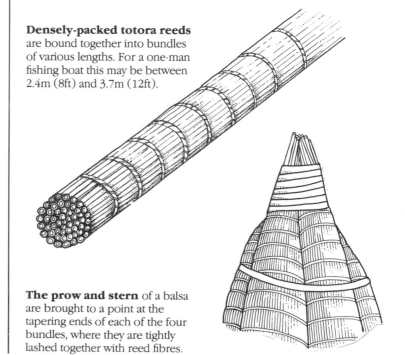

Densely-packed totora reeds are bound together into bundles of various lengths. For a one-man fishing boat this may be between 2.4m (8ft) and 3.7m (12ft).

The prow and stern of a balsa are brought to a point at the tapering ends of each of the four bundles, where they are tightly lashed together with reed fibres.

Uru and Aymara Indians harvest the bountiful supply of totora reeds from the shallows of Lake Titicaca, especially from Puno Bay on the west coast. The Indians spread out the long, tubular stems to dry in the sun, as farmers elsewhere in the world might leave grass to turn into hay. When dry, the totora reeds become crisp and fragile; but after a soaking in water they assume the pliancy and resilience of rope.

Tests carried out in 1956 showed that, after more than a year's constant immersion in water, a boat made from totora reed was left undamaged by waterlogging, or by attack from marine fauna and flora, such as barnacles and algae.

Floating islands fashioned from totora reeds provide a home for some Uru and Aymara Indians. They build boats and houses, and cook food on the islands' spongy buoyant surface. Each island is composed of tightly bound bundles of totora, and measures around 2m (6.6ft) thick. The Indians also eat totora pulp, after pulling the reeds up by the roots and peeling the stems.

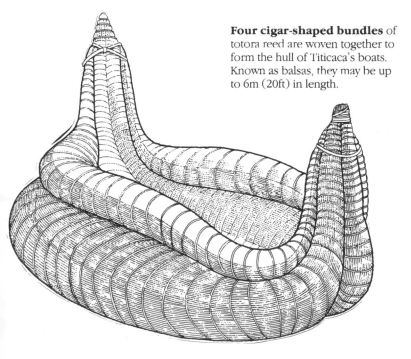

Four cigar-shaped bundles of totora reed are woven together to form the hull of Titicaca's boats. Known as balsas, they may be up to 6m (20ft) in length.

IGUAÇU FALLS

A multitude of cascades deep in the jungle

The ruptured edge of southern Brazil's enormous Paraná Plateau creates the natural setting for a number of waterfalls. The Iguaçu Falls are unanimously acclaimed as the most spectacular. Swiss botanist Robert Chodat (1865–1934) eloquently described their imposing grandeur: 'When we stand at the foot of this world of cascades and, raising our eyes, see, 269 feet above us, the horizon filled with a line of waters, this awesome spectacle of an ocean pouring into an abyss is almost frightening.'

Nearly 275 individual falls, each separated from the next by rocky islands covered in trees, make up Iguaçu's mighty cascade. Some falls tumble unbroken from the rim, which is almost 4km (2.5mi) long, to the gorge yawning 82m (269ft) below. Many others descend the precipice in a series of short cascades as their waters spill from ledge to ledge. No other falls of comparable size are divided into so many different channels. The dissection is thought to be brought about by the geology of the plateau, which is formed of solidified lava and tough volcanic rocks, such as basalt. These rocks are not easily eroded; thus they force water around islands and into channels.

The rise and fall of the Rio Iguaçu's waters depend almost entirely on the seasonal rains falling into its drainage basin. During the height of the rainy season, from November to March, the river swells to a tremendous flood when as many as 12,768,750 litres (2,812,500 gallons) of water flow over the Iguaçu Falls every second; this figure is equivalent to more than six Olympic-sized swimming pools, and more than the total volume of the dome of the Capitol in Washington, D.C., USA.

'After seeing Iguaçu Falls,' remarked Eleanor Roosevelt (1884–1962), wife of former US President Franklin D. Roosevelt, 'it makes our Niagara Falls look like a kitchen faucet.' Compared to

The Iguaçu Falls lie on the border between Brazil and Argentina, and only 19km (12mi) east of the Paraguayan border (*left*). The Rio Iguaçu rises near the Brazilian city of Curitiba in the Serra do Mar mountains and is fed by 30 tributaries as it flows the 1,300km (800mi) to the Iguaçu Falls. As they approach the falls, the river waters spread over a wide area (*right*) before plunging over the rugged precipice in a multitude of individual cascades.

Niagara Falls, Iguaçu is four times as wide, half as high again and carries seven times as much water when in full spate.

During the dry season between April and October the water cascading over the Iguaçu Falls decreases dramatically as the river dwindles to a fraction of its maximum size. In 1978, the Iguaçu dried up completely. The falls became a line of rocky cliffs and remained so for four weeks before a trickle of water returned. Such dry spells, however, occur on average every 40 years.

The waters of the Rio Iguaçu plunge into a narrow gorge known as the Garganta del Diablo, the 'Devil's throat', then wind their way to join the Rio Paraná around 22km (14mi) to the south. After the Amazon and the Orinoco, this is South America's third greatest river. It rises far to the north and flows out into the South Atlantic Ocean via the Rio de la Plata.

In 1541, the Spanish explorer Alvar Núñez de Vaca became the first European to discover the Iguaçu Falls. Inspired by the piety of his generation, de Vaca named them Salto de Santa María, the Falls of St Mary. But the title did not stick; the falls soon reverted to the local Guirani name of Iguaçu, which means 'great water'.

In the following two centuries, Spanish Jesuit priests explored and investigated the falls and the river system around them. They established missions to improve the lives of the Guirani Indians and to introduce them to ideas about democracy, trade and civil rights. The Jesuits protected the Indians from Spanish and Portuguese landowners who needed slaves to work their plantations near the coast. After many years of religious and legal battles in Spain, as well as armed conflict in the jungles themselves, the secular landowners triumphed over the Jesuits and had them expelled from South America in 1767.

Natural history of the falls

In his eulogy to Iguaçu Falls, Swiss botanist Robert Chodat also portrays the rich flora thriving there: '. . . an exuberant, almost tropical vegetation, the fronds of great ferns, the shafts of bamboos, the graceful trunks of palm trees, and a thousand species of trees, their crowns bending over the gulf adorned with mosses, pink begonias, golden orchids, brilliant bromeliads and lianas with trumpet flowers. . . .'

The national parks which Brazil and Argentina established early in the 20th century on their respective sides of the falls protect the rich tropical and subtropical wildlife. Birds such as tinamous and parrots haunt the trees, while swifts nest in the craggy outcrops of the falls and swoop low over the river to feed on the large swarms of insects. Ocelots and jaguars roam the forest wilderness, as do tapirs, three species of deer and two species of peccary.

On the rocky islands dividing the river water into channels grow a variety of trees, such as the cedar (*Cedrella fissilis*), and trumpet vines, such as two species of the lapacho (*Tecoma ipe* and *T. ochracea*). A number of unusual aquatic herbs from the family Podostemaceae thrive on the ledges of the falls; these are flowering plants although they resemble water lichens or mosses. All the members of this family grow only in running water: some prefer to live in the midst of the waterfall's spray, others hide under a ledge or else brave the full might of the rushing water.

All these plants, which rarely exceed 10cm (4in) in height, fix themselves firmly to the rock by means of suckers. When the water level falls in the dry season, they flower and immediately release pollen which fertilizes the ovaries of adjacent plants. Fruits ripen within a few days, drop on to the nearby rock and affix themselves at once. When the water level rises again, the seeds germinate and produce new plants.

The forest-dwelling Brazilian tapir (*Tapirus terrestris*) is a common visitor to the waters around the Iguaçu Falls. A good swimmer and diver, this stocky, dark brown creature is also surefooted on land, no matter how rugged the terrain. At night, the tapir browses on leaves, buds, fruit and water plants.

The solitary jaguar (*Panthera onca*) is the largest cat in South America. Using its thick, powerful limbs it climbs trees and swims across rivers. Jaguars prey on animals such as peccaries and capybaras that drink at the water's edge, but they will also eat fish, caimans, deer, otters, turtles and ground-living birds.

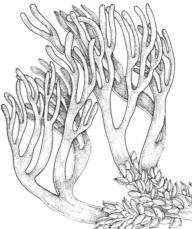

Some aquatic herbs resembling mosses and lichens grow only in fast-flowing water such as waterfalls. The flowers of these plants, including *Dicraeia algiformis* (*left*), have no petals, and bloom at the end of the rainy season. All are members of the family Podostemaceae. Seeds dispersed in the subsequent dry season do not germinate until they are covered by water.

Myriad islands break up the flow of the Iguaçu River prior to its plunge over the edge of the Paraná Plateau. Much of the river's water cascades over Union Falls and San Martín Falls, which lie at opposite ends of Iguaçu's precipice. Between these giant falls, a multitude of smaller cataracts, such as the Three Musketeers and the Belgrano, spill over ledges and terraces carpeted with dense vegetation. Birds, such as swifts, build their nests on craggy outcrops between the falls and swoop low across the river above to feed on swarms of insects.

SURTSEY

An island born from beneath the sea

Before dawn broke on November 14, 1963, a fishing vessel, the *Isleifur II*, moved slowly through the sea west of Geirfuglasker in Iceland. The early morning was quiet and peaceful as the crew put down their long cod lines. Suddenly a large wave struck the boat, pitching it steeply. On recovering their balance the fishermen sighted a tall column of smoke to the southwest.

The captain of the *Isleifur II* thought a ship was on fire: he radioed the coastguard and headed toward the smoke to render assistance. But as the fishermen neared the fire, the truth dawned. It was not a burning ship but a volcanic eruption bursting through the sea. The captain immediately altered course and sped for home. Clouds of steam billowed from the water. Periodic explosions hurled rocks high into the air. Within three hours of the first eruption, the column of smoke and ash had climbed to around 3,660m (12,000ft).

The underwater volcano had ripped open about 2.6sq.km (1sq.mi) of sea bed, which lay only 130m (426ft) below the surface of the water. Friction between dust particles generated fantastic displays of lightning. Explosions churned the sea into a frenzy. Massive waves threatened to overwhelm passing boats. The sudden conversion of countless gallons of water into steam triggered further undersea explosions. So violent were these that the orange-hot magma from the Earth's core was transformed into an enormous shower of fine particles.

Journalists and scientists, who had flocked from all over the world to witness the birth of an island, were pelted with cinders, pumice and fine ash, a mixture known as tephra. The eruption column, which occasionally rose as much as 15,240m (50,000ft), could be seen from Reykjavík, Iceland's capital, around 120km (75mi) to the northwest.

The island of Surtsey lies about 20km (12mi) southwest of Heimaey in the Westman Islands and around 35km (22mi) from the shores of Iceland (*left*). When seen from above, the island resembles the outline of a rodent's skull; the conical 'snout' of the island (*right*) points northward toward Iceland. Two rounded craters at the centre of the island are the remnants of the volcanoes that bore tephra and then lava from beneath the sea.

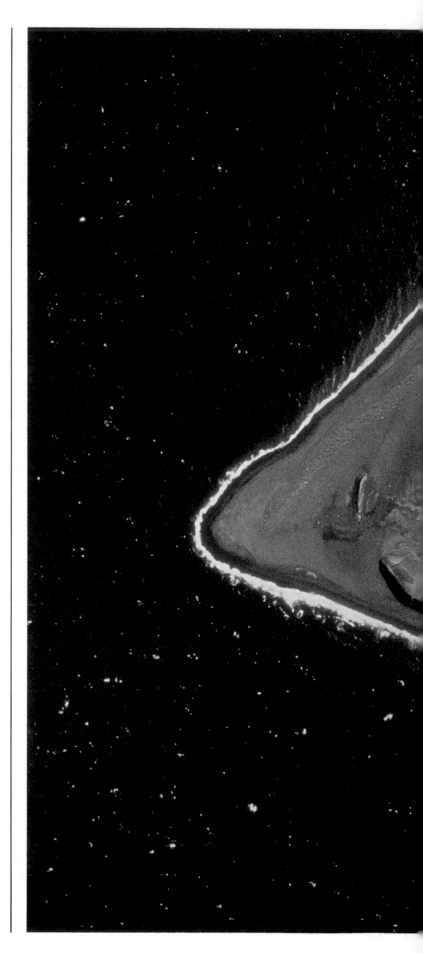

After two days of observation, scientists were able to discern an object in the heart of the dense, billowing cloud. A huge ridge of rock began to take shape. Before long an island, about 40m (130ft) in height and 550m (1,800ft) in length, was clearly visible. Four weeks later, the volcano appeared to have subsided. But as a small group of journalists from the French magazine *Paris-Match* stepped cautiously ashore, they were bombarded by ash and pumice, and forced to withdraw.

The Government of Iceland's Place Name Committee called the volcano Surtur, after a fire giant of Scandinavia's pagan mythology. According to these myths, the end of the world, Ragnarok or the 'twilight of the gods', would be heralded by several terrible events. One of these was the coming of Surtur, who would ride across the world striking indiscriminately with his sword of fire.

The formation of the island

By the end of January 1964, the growing island, now named Surtsey, rose about 150m (500ft) above the sea and covered about 2.6sq.km (1sq.mi), an area roughly half the size of New York City's Central Park. Scientists suspected that it would exist only for a short period, since the pumice and ash of the tephra, the island's chief constituent, would be unable to resist the battering of the waves and wind.

But in February 1964, the northwest face of Surtsey erupted, and a second volcano poured out large streams of lava. As the lava spread outward from the cone of this volcano, named Surtur Junior by watching scientists, it flowed over the tephra and solidified into a tough shield at the island's northern edge. Eventually lava flowed out of Surtur itself and mingled with the tephra to form a substance which could withstand the fiercest storms. It was obvious that Surtsey would be in place for a long time.

Surtur is one of many volcanoes to erupt from the Mid-Atlantic Ridge. This mountain chain, most of it below sea level, runs for more than 16,000km (10,000mi) down the centre of the Atlantic Ocean from Jan Mayen Island in the north to the island of Bouvet in the south. Iceland is the largest above-sea island of the ridge and all of its satellite islands have, like Surtsey, appeared from beneath the ocean.

The eruptions on Surtsey ceased altogether in 1967. The bulk of the island is a curved ridge marking the edge of Surtur's rugged crater. The jagged crest of this crater, measuring almost 1.6km (1mi) across, rises 171m (561ft) above the surrounding sea. A smaller crater raises its craggy outline inside the larger one. Masses of solidified lava lie scattered in jumbled heaps, while sandy deposits form beaches along the coast.

The formation of the brand new island of Surtsey has provided modern scientists with a unique opportunity to study the colonization of sterile territory by plants and animals. From the moment of its creation, Surtsey became a laboratory. Visitors were forbidden lest they inadvertently brought seeds or spores with them in their clothes or on their footwear. Only teams of scientists wearing sterile clothing were permitted to land on the island to investigate the volcano and any scanty signs of life.

The island's shore was first colonized by bacteria, moulds, seaweeds and green algae. After three years mosses took hold on the lava, followed five years later by lichens. Enough seeds have arrived, either borne on the wind or else airlifted by seabirds, to produce a scattering of grasses and sedges. In 1970, fulmars (*Fulmaris glacialis*) and black guillemots (*Cepphus grylle*), started nesting in cliffs on the western side of the island. In the absence of eggstealing mammals, such as rats, these and other sea birds are able to breed in perfect safety.

A submarine volcano explodes violently from the crest of the Mid-Atlantic Ridge and begins the process that will create the island of Surtsey (*above*). Billowing clouds of cinders, ash and pumice darken the sky off the southwest coast of Iceland. At the time, observers were pelted with lava bombs – deadly missiles of liquid rock catapulted high into the air during the eruption. As each missile fell, its surface cooled into a craggy skin which enveloped a ball of hot lava.

The mayweed (*Matricaria maritima*) became one of the first flowering plants to appear on the island of Surtsey, although it did not take root successfully until 1972. A common plant elsewhere in Iceland, the mayweed grew from bird-borne seed in the sand-covered lava at Surtsey's southern tip.

The fulmar (*Fulmarus glacialis*) was, together with the guillemot, the first bird species to colonize Surtsey. In 1970, a fulmar built a nest on a ledge some 10m (33ft) above the sea on the island's western cliffs. The next year, there were 10 fulmar nests and, by the 1980s, a breeding colony had been fully established.

The island of Little Surtsey, or Syrtlingur, was born in 1965, some 600m (1,969ft) off Surtsey's eastern shore. Submarine eruptions, which threw up water spouts and pumice were followed by violent explosions that pushed a tiny island of tephra above the waves. After a few months, Little Surtsey drew to a height of 70m (230ft) and covered an area of 14 hectares (35 acres). Without a protective crust of lava, Little Surtsey was soon washed away and remains only as shallow shoals beneath the water.

STROKKUR

Fountain in the land of ice and fire

In AD 860, a Viking longship, commanded by the Norwegian pirate Naddod, set sail for the Faroe Islands in the North Sea. But it was blown off course by a tremendous storm. Naddod, thus forced to land on a previously unknown island, was able to repair his ship and return home to Norway. The tales he told enticed fellow Norwegian Floki Vildergarson to explore the new island. On seeing the huge icebergs floating in the fjords he named the country 'iceland'. Later Vikings believed the island was the entrance to the underworld: contrasts between volcanoes and glaciers, fire and ice were exactly as those described in the tales of the Norse gods.

Iceland, Europe's second largest island after Britain, was formed from submarine eruptions in the Mid-Atlantic Ridge. The island remains a land of contrasting ice and fire: it boasts the largest glacier, Vatnajökull, between the Arctic and Antarctic circles and in the last 2,000 years it has suffered more than 125 eruptions from 30 different volcanoes.

The earliest recorded visitor to Iceland was the Irish monk Brendan the Voyager who, nearly three centuries before Naddod, travelled with several monastic brothers in search of solitude on remote islands. Around AD 575, they came within sight of Iceland. A medieval Latin text, *Navigatio Sancti Brendani*, records what he saw: 'There came into view a large and misty mountain in the ocean . . . with misty clouds above it, and a great smoke issuing from its summit . . . Then they saw the peak of the mountain unclouded and shooting up flames into the sky which drew it back again to itself so that the mountain was a burning pyre . . .'

Brendan and his fellow monks probably witnessed an eruption of Mount Hekla, a volcano 1,491m (4,892ft) high in the southwest of Iceland. To the northwest of this mountain, lies the

Strokkur is located in the geothermal region beside the Hvítá River, midway between Mt. Hekla and Thingvellir in the southwest of Iceland (*left*). The country's capital, Reykjavík, lies roughly 80km (50mi) to the west. From a quiet, steam-covered pool, Strokkur suddenly and momentarily swells into a dramatic dome of scalding water (*right*). On eruption, the plume of Strokkur's fountain reaches heights of 30m (100ft), its white cloud of steam visible to the naked eye from a distance of more than 5km (3mi).

valley of the Hvítá River. Here, grouped together in one of Iceland's many geothermal regions, are steaming pools of scalding water and bubbling mud. This is the home of the most famous hot spring of all: the Stori Geysir, which in Icelandic means 'the great gusher'. From its name, the word geyser has come to describe all similar geothermal fountains.

In the 1980s, the Stori Geysir is little more than a still, blue pool of water, some 10m (33ft) across and surrounded by a wall of silica and the calcareous rock travertine built up over many centuries. The waters of the Stori Geysir, once the largest of a group of around 100 geysers, which were already celebrated in the 12th century, reached a height of 70m (230ft). In 1810, it erupted at 30-hour intervals: by 1815 it played every 6 hours for up to 15 minutes, pushing up several jets of water at once.

In 1916, the great gusher suddenly and inexplicably stopped. Then, in 1935, it mysteriously started up again, hurling its plumes to the same height as before. In the 1980s it rarely plays although it can be encouraged to do so by the application of large amounts of soap to its waters. This reduces the surface tension of the water and eases its flow from the hot subterranean chambers. Another way of encouraging geysers to spout is to block the tube through which the water rises, thereby cutting down heat loss and allowing the water to boil more quickly.

Heir to the great gusher

The geothermal fountain of Strokkur has taken on the mantle of Iceland's greatest geyser. At first sight, it appears as a quiet, shallow pond of brilliantly clear water – only the steam hints at Strokkur's secret. The initial sign of approaching activity is a fluctuation in the pool's water level. The waters heave and sigh with increasing speed. Suddenly, from the heart of the pool, an astonishing dome of crystal clear water wells upward.

The Strokkur dome exists only for an instant before the pool erupts. The smooth outline of the water dome bursts apart and a gigantic shaft of steam and boiling water shoots up to a height of 30m (100ft). The scalding column hangs in the air for several moments before dispersing down the valley on the breeze. Shifting and changing shape in ghostlike fashion, the white cloud floats away and disappears into thin air.

The water hurled high by the eruption drops turbulently back into the Strokkur pool. Eventually, the water surface settles down to become placid and steam-covered once again. But Strokkur, which means 'the churn' in Icelandic, does not remain calm for long. On average its eruption sequence is repeated about seven times an hour, day and night.

Icelanders recognized the benefits of using the heat from deep inside the Earth in the early part of the 20th century. By 1928, some homes in Reykjavík, Iceland's capital, were heated with natural hot water. In 1942, engineers completed an extensive system of pipes and pumping stations which brought natural hot water to massive tanks around Reykjavík's hills. The water sources were carefully chosen so as not to disrupt the playing of Strokkur and other geysers. By the 1980s, nearly every home in Reykjavík was plumbed into the geothermal hot water system. Tap water is warmed by means of heat exchangers which transfer heat from the primary to the domestic supply.

Not all the energy is channelled into industrial and domestic usage. Where the Hvítá River valley opens out into a plain lies the town of Hveragerði, 'the garden of hot springs'. Here, only 250km (156mi) south of the Arctic Circle and in glasshouses heated by geothermal waters, tropical houseplants, and fruit and vegetables, such as bananas, cucumbers and gherkins, are grown.

Geothermal phenomena such as the steaming rocks beside Kleifarvatan, a lake to the south of Reykjavík, occur in many parts of Iceland. Geologically, the island is still comparatively young, and molten magma runs near the Earth's surface. When it heats the subterranean water to boiling point, the resulting steam is forced out through cracks and fissures in the ground.

Steam wells drilled at Krafla in the northeast of Iceland typify the country's increasing development of its tremendous resource of geothermal power. By 1984, geothermal energy heated some 80 per cent of Icelandic homes, and satisfied 37 per cent of the island's energy needs. Yet only 5 per cent of the potential power that could be economically exploited was being tapped.

Strokkur's geyser is a remarkably reliable performer. It erupts on average at intervals of nine minutes around the clock. The timing of these bursts is fairly constant over the course of a day, but will vary from week to week – from just a few minutes apart up to 20 minutes or more. The hot water gushes up through an opening 2m (6.5ft) in diameter into the pool, which is some 10m (33ft) across.

LAS MARISMAS

Bustling bird crossroads and marshland haven

Twenty-five centuries ago, Phoenician war galleys travelling to and from the port of Gades, the modern Cadiz, traversed the large bay at the mouth of the River Guadalquivir in southwest Spain. Twelve centuries later the bay had been transformed into an enormous salt lake. By the 20th century, the lake had become a trackless marshland which, within the next millenium, will probably become dry land.

The combined bulk of silt transported by the Guadalquivir and the power of strong sea currents have wrought this transformation of the landscape. Sand swept along the coast from the west gradually formed a sandbank across the bay, impounding a shallow sea. The river's cargo of silt and mud, brought down from the Sierra Morena and other inland mountains, drove out the salt water and created shallow lagoons in its place. In the 1980s, this barrier, known as Arenas Gordas or 'fat sands', stretches for 70km (45mi). It is only breached by the river at its southern end. The shifting sand dunes can be as broad as 13km (8mi) and are dotted with pines, tough grasses and scrub.

Las Marismas, which in Spanish means 'the marshes', covers an area of 1,150sq.km (450sq.mi). Biologically diverse habitats, such as dune heath, savanna and the remains of cork-oak forests, border the brackish marshes. Together they provide sanctuary for a multitude of birds, both resident and migrant, as well as several beleaguered species of mammal.

By the 13th century, Las Marismas had become a royal *coto*, or hunting reserve. In 1294, King Sancho IV donated the land to Alonso Perez de Guzman as a reward for his part in preventing the Moors from taking the town of Tarifa, near Gibraltar. At the same time, the king bestowed Guzman with the title of Duke of Medina-Sindonia. Early in the 17th century, the seventh duke

Las Marismas is located 24km (15mi) to the southwest of Seville where it forms part of the delta of the River Guadalquivir as it flows into the Gulf of Cadiz (*left*). Formerly a royal hunting ground, the flat plain of Las Marismas harbours Europe's richest congregation of wildlife. Some parts of the swampy marshland (*right*) are flooded seasonally, while others are permanently under water. At the margins, there are a variety of habitats – pinewood islands, dune heath, pastureland and ancient cork-oak woodland.

built a lodge amid the marshlands as a retreat for his wife Doña Ana. This lodge, known as El Palacio, remains the only building of note in the area.

The absence of any development in the marshland has meant that its unique ecology has been largely preserved. However, extensive coastal resorts on either side have, since the 1950s, posed a threat to the ecological balance of Las Marismas. Only the intervention of eminent European biologists, led by Spanish scientist Dr José Valverde and supported by the newly-formed World Wildlife Fund, ensured the survival of this wetland sanctuary. In 1964, some 6,500 hectares (16,055 acres) were set aside as the Coto Doñana nature reserve. In 1969, it was merged with the adjacent Guadiamar reserve; the total area of 35,000 hectares (86,450 acres) was, at the same time, given national park status.

Unrivalled haven for wildlife

The Coto Doñana National Park has become the last refuge for several endangered species of European birds, such as the black vulture, and mammals, such as the Spanish lynx. Foremost of these is the Spanish imperial eagle (*Aquila heliaca*). Distinguished by white shoulders and wing edges, this majestic bird of prey hunts small mammals, including rabbits and hares. In 1977, estimates put the eagle's breeding population at 60 pairs.

Las Marismas undergoes radical seasonal changes. Heavy rains sweeping across southern Spain in late autumn inundate the western slopes of the Sierra Nevada and the plain of Andalucia. The silt-laden water drains into the Guadalquivir which, in turn, floods Las Marismas, often to a depth of 60cm (2ft). The wetland environment provides a welcome stopping-off point for a multitude of waterfowl migrating south for the winter, and also a winter home for many ducks and geese returning from their summer breeding grounds in northern Europe and Scandinavia. For a few weeks, Las Marismas is turned into a noisy, bustling crossroads for a million avian migrants.

In spring, when the mountain snows melt, fresh water descends to the marshes, while the rise in temperature stimulates plants to grow. The dormant waterscape comes alive. Thick stands of grasses, reeds and rushes spring up on elevated islands, or vetas, and help conceal an enormous number of bird nests. In early May, swarms of midges, mosquitoes and dragonflies emerge from their pupal stages and lure birds to the neighbourhood. The fluttering wings and mating calls of hundreds of thousands of birds bring cacophony to the marshland.

A clamouring congregation of 173 different bird species, many of which breed nowhere else in Europe, gathers together to feed and mate: all but one of Europe's nine species of heron, flamingoes, black-winged stilts, turnstones and plovers. The breeding community also includes rare birds, such as the ruddy shelduck, the crested coot and the marbled teal. Red kites and black vultures wheel over Las Marismas in search of lizards and small mammals. Egrets and spoonbills form weighty nesting colonies in ancient cork oaks, while in the dense thickets of bramble and jaguarzo shrubs the Spanish lynx makes its lair.

Within a few weeks the Guadalquivir ceases to top up the marshes; the sun's heat intensifies and evaporates the water. In dwindling water channels rushes and grasses grow in ever thickening clusters while green algae accumulate in huge clumps. Large mammals appear in increasing numbers. By June, herds of red deer and fallow deer splash through the water to reach succulent grasses. Wild boar shuffle across drier ground. By early August, the relentless sun has scorched Las Marismas to a cracked mosaic of concrete-hard mud.

Dedicated ornithologists have discovered that the best views of Las Marismas' unparalleled bird population are obtained the old-fashioned way. Flat-bottomed duckboats are drawn by patient Andalusian horses with tow ropes knotted to their tails. The horses splash knee-deep through the spring waters. Until the late 1960s, there was no other access to Coto Doñana National Park, although now an asphalt road skirts the edge of the marsh. It finishes at the 17th-century Palacio de Doñana, which has been converted from a hunting lodge to a biological research station studying the park's wildlife.

Reeds poke up through a glutinous mass of algae clotting the surface of Las Marismas' brackish waters – an almost primeval habitat in which countless forms of wild life flourish. In spring, the wetlands are still flooded with snowmelt carried down from the Sierra Morena, but from May onward the inexorable parching of the swamps begins again.

The Spanish imperial eagle (*Aquila heliaca adalberti*) is distinguished from other races of imperial eagle by the greater extent of its white plumage.

The red kite (*Milvus milvus*) (*above*) is among the larger hawks at 61–66cm (24–26in) in length. A colourful predator, it nests in Las Marismas' cork oaks.

SKYE

Island sculpted by volcanic upheaval

The rugged backbone of the Black Cuillin Hills dominates the view from almost every vantage point on Scotland's Isle of Skye. English journalist H.V.Morton (1892–1979) amply described the earthscape in his book *In Search of Scotland.* 'Imagine Wagner's 'Ride of the Valkyries' frozen in stone and hung up like a colossal screen against the sky. It seems as if Nature when she hurled the Coolins up . . . said: "I will make mountains which shall be the essence of all that can be terrible in mountains".'

The narrow strait of Loch Alsh separates the Isle of Skye from the Scottish mainland. Dozens of rugged peninsulas define the island's coastline, which is so convoluted that it extends for almost 1,600km (1,000mi). In places the coast presents a gentle prospect, as at Portree, the island's major town, where white-washed fishermen's houses stand beside a 19th-century harbour. Mostly the island is ringed by precipitous hills and isolated promontories. At Dunvegan Head in the northwest, a sheer cliff rises 313m (1,028ft) from the waves to a rounded, grassy summit.

According to legend, the island was once flat moorland and inhabited by Cailleach Bhur, the goddess of winter. She had enslaved a beautiful maiden, the sweetheart of spring, who appealed to the sun for help. In response, the angry sun hurled his burning spear at Cailleach Bhur as she strode across Skye. But he missed his mark and blistered the landscape which erupted into a range of hills – the Cuillins. Skye's inhabitants often relate this tale to explain the curious fact that these hills are rarely snow-covered, even in winter. When surrounding glens and hills are draped by a glistening white blanket, the Black Cuillins stand stark and dark against the sky.

Twenty peaks feature in the Black Cuillins, 15 of them more than 914m (3,000ft) in height. The tallest, Sgur Alasdair, stands

Skye is located off Scotland's northwest coast, some 184km (115mi) from Glasgow (*left*). The deeply indented island measures 77km (48mi) in length and 38km (24mi) at its widest. The largest island in Scotland's Inner Hebrides, Skye covers an area of 1,740sq.km (672sq.mi), about half the size of the state of Rhode Island in the USA. Skye's landscape is dominated by the spectacular Black Cuillin Hills, particularly from vantage points such as Tarskavaig (*right*) in the southeast of the island.

1,009m (3,309ft) high. The peaks came into existence some 50 million years ago when vast amounts of lava poured out of vents in the Earth's crust. The hills are composed of gabbro, a hard rock derived from volcanic material that cooled slowly underground. Subterranean activity thrust them upward where their brittle substance was fashioned by the power of glaciers.

The Red Cuillin Hills, some 16km (10mi) to the east are composed of granite – the same volcanic material as gabbro, but softer because it cooled rapidly above ground. These hills are more rounded than the Black Cuillins, which resemble shards of broken crockery. The granite scree slopes are tinged pink and at sunset on a clear day appear like rivers of pale blood flowing down the mountainsides.

In the north of the island stand the unusual cliffs and ravines of the Quirang, a name which in Gaelic means 'the rounded fold'. Around 19km (12mi) to the south a rocky pinnacle, 50m (165ft) tall and known as the Old Man of Storr, points skyward. Both these formations had their beginnings 10,000 years ago when the ice age glaciers retreated and left behind hard basalt rocks perched on soft clay slopes. As the clay collapsed and shifted, so the basalt rocks slipped or were stranded in curious positions.

The numerous lochs around the island are another ice age legacy. Valleys gouged out by glaciers were filled by seawater once the ice fields had retreated. Many of the lochs, such as Loch Coriusk at the foot of the Black Cuillin Hills, open like fjords to the sea. Scottish author Sir Walter Scott (1771–1832) described the misty mood of the loch: 'The murky vapours which enveloped the mountain ridges obliged us by assuming a thousand varied shapes, changing their drapery into all sorts of forms, and sometimes clearing off altogether.'

The island's flora and fauna

Little of Skye's landscape can support the barley, oats and other crops grown elsewhere in Scotland. Areas of fertility, and therefore human settlement, occur only where either sandstone, limestone or clay covers the volcanic rock. In general, large flocks of sheep, such as the hardy Scottish blackface and the cheviot, roam the moor and grasslands.

Despite the patchy fertility of its soil, the Isle of Skye has a rich flora: studies have revealed 589 flowering plant and fern species, 370 mosses, 181 liverworts and 154 lichens. Rare species of flowering plant include the red broomrape (*Orobanche alba*) and the Iceland purslane (*Koenigia islandica*). Grouse shooting takes place in carefully managed heather moors. A mature heather (*Erica spp.*) is a tough, woody plant with few young shoots to provide food for grouse. Because of this, old plants are burned off every 12 years to allow seeds to germinate and fresh shoots to grow.

Thousands of seabirds, including the Arctic tern and the great skua, nest on the island's ragged coast and shore. The puffin (*Fratercula arctica*), which nests on precarious cliffs, can catch small fish in prodigious numbers. It is not unusual for a puffin to return to its nest with as many as 14 sprats in its mouth. Yet Skye's puffin population is on the decline, due largely to the attacks of lesser black-backed gulls which rob the puffins of their catch before they can reach their nests to feed their young.

Perhaps the most spectacular sight of all is the golden eagle (*Aquila chrysaetos*), which haunts the inaccessible reaches of the island's cliffs and peaks, especially the Black Cuillin Hills. Around 80cm (32in) long, this magnificent bird of prey is equipped with formidable curved talons which it uses to seize rabbits, hares and the occasional grouse.

The Quirang, on Skye's northern Trotternish peninsula, is a cuplike configuration of oddly isolated peaks composed of volcanic basalt. These survived the retreat of ice-age glaciers, which, upon melting, swept away the soft clays that originally supported them. Individual formations in this intriguing rocky girdle include the Table, the Prison and the Needle, which stands 36m (120ft) high.

Loch Coriusk nestles in the heart of the Black Cuillin Hills some 22km (14mi) south of Portree. The land around the loch remains the main haunt of Skye's red deer, Britain's largest wild mammal. The basin floor of Coriusk, which in Gaelic means 'cauldron of water', was excavated by a glacier and lies around 30m (100ft) below sea level.

Crofters gather in the hay harvest from Skye's comparatively small areas of arable land. The agricultural emphasis is on grazing – mainly Scottish blackface sheep, and some beef cattle. But there is still a need to grow winter fodder, and the fertile pockets of workable soil – notably in northern Trotternish, Broadford and the west of Slapin – are put to optimum use.

GIANT'S CAUSEWAY

Ireland's legendary stepping stones

According to legend, the bad-tempered Irish giant Finn MacCool built a road across the waves to reach his enemy Finn Gall, who lived on Scotland's Isle of Staffa. MacCool gathered together a multitude of long, stone stakes and hammered them into the seabed next to one another. Before challenging Finn Gall to a duel, the Irish giant returned home to rest. Meanwhile, Finn Gall crossed to Ireland and was deceived into thinking that the sleeping giant was MacCool's baby son. Terrified at the possible size of the father, Finn Gall fled home to Staffa, destroying the causeway as he went.

The stone columns of this Giant's Causeway are stacked on the coast of Northern Ireland's County Antrim. While few people take seriously the fanciful tale of the mighty giants, the regularity of the columns does lend to the illusion that they were manmade. To quote the 19th-century Irish poet W.H. Drummond: 'A far projecting, firm, basaltic way / Of clustering columns wedged in dense array; / . . . reason pauses, doubtful if it stand / The work of mortal or immortal hand.'

The spectacular array of columns spreads along 275m (900ft) of coast and reaches as far as 150m (500ft) into the sea. An anonymous reckoner from the 1930s decided there were around 40,000 columns present – no one knows how the count was made and yet the figure has never seriously been challenged. Most of the columns stand no higher than 6m (20ft), although some, such as the Giant's Organ (so named because of its resemblance to a church organ), reach some 12m (39ft).

Each individual column, shaped into a regular polygon, measures between 38cm (15in) and 50cm (20in) across. Most are six-sided, while others may have four, five or as many as ten faces. When viewed from above, the causeway resembles a street

The Giant's Causeway lies on the north coast of County Antrim in Northern Ireland, some 80km (50mi) northwest of Belfast, the country's capital (*left*). The highlight of a dramatic shoreline, the assemblage of basalt columns (*right*) covers an area of around 2 hectares (5 acres). Many of the ancient, hexagonal columns lie broken and shattered on the beach, while others have been swallowed by the sea or buried in Antrim's soil.

with regular paving stones – the columns fit together so exactly that it is difficult to insert a knifeblade between them.

Drawings and sketches made in the 18th century by the Dublin Society and Britain's Royal Society alerted the scientific world to this remarkable phenomenon. The paintings, commissioned by the Earl-Bishop of Derry, Frederick Hervey, in the late 18th century, brought the causeway to the attention of the public.

To the Romantic Movement, which flourished at the start of the 19th century, the Giant's Causeway was the living proof of all they believed in. One leading exponent declared that the array of stone columns was 'the temple and altar of Nature, devised by her own ingenuity, and executed with a symmetry and grace, a grandeur and a boldness which Nature alone could accomplish.'

Formation of the columns

During the first half of the 19th century, the Giant's Causeway became the centre of a furious geological debate. On the one side were Vulcanites who believed that volcanoes were as old as the Earth and the basalt columns had been formed from solidified volcanic lava. On the other side were the Neptunites, fundamentalists led by Irishman Richard Kirwan who claimed that 'sound geology graduates into religion'. The Neptunites held that volcanoes were geologically recent phenomena, and that the basalt rocks, because they were old, must have been formed by the precipitation of minerals from seawater.

The Vulcanites won the argument, and today geologists of repute consider the Giant's Causeway to be volcanic in origin. Around 50 million years ago, much of Northern Ireland and western Scotland became volcanically active. Vents in the Earth's crust, such as those at Slemish Mountain, opened up time and again, pouring lava over the landscape to a depth of almost 160m (525ft). When the lava cooled quickly, it solidified to form basalt, a tough, erosion-resistant rock. At Slemish, the basalt is piled up to a height of 441m (1,457ft) above sea level.

At the Giant's Causeway, however, the lava cooled slowly and steadily. As the upper levels lost their heat first, they shrank and cracked into regular patterns, much as mud cracks when it dries. The surface fissures gradually extended downward, splitting the entire mass of basalt rock into an array of upright columns of stone. Softer surrounding rock has been eroded away at the coast, exposing the basalt columns to the air. The formation is thought to extend inland, beyond the spectacular line of cliffs known as the Giant's Cuffs and beneath the green landscape of County Antrim.

The flat-topped Isle of Staffa, lying 120km (75mi) to the north of the Giant's Causeway, is also renowned for its hexagonal columns. Almost entirely composed of basalt, the island is ringed by cliffs of sheer, fluted columns and topped by a mop of spongy rock. Sir Joseph Banks (1743–1820), the English naturalist who accompanied Captain Cook on his 1768 voyage of exploration to the South Seas, brought the island to the public's attention in 1772. 'Compared to this,' he exclaimed, 'what are the cathedrals or the places built by man . . . mere models or playthings.'

A huge gothic cave penetrating some 60m (200ft) into the island was named Fingal's Cave by Sir Joseph Banks after the legendary giant, Finn Gall. At low tide, the cave roof is about 18m (60ft) above the water. At high tide, or when the sea is whipped up by Atlantic storms, the water forced into the cave compresses the air and generates a rhythmic 'singing' sound. The young German composer Felix Mendelssohn (1809–1847), who visited the Isle of Staffa in 1829, was so captivated by this sound he penned his Fingal's Cave overture the following year.

Thousands of basalt columns formed from solidified volcanic lava crowd Northern Ireland's shore northeast of Coleraine. The columns appear to be organized into three natural platforms, known as Little Causeway, Middle Causeway and Grand Causeway. Within these platforms, groups of columns have been assigned such fanciful names as the Wishing Chair, the Chimney Tops and the Giant's Organ.

The remarkable geometry of the causeway's columns resulted from the slow but even rate at which the volcanic lava cooled into basalt. As the cooling progressed, the basalt assumed regular, polygonal shapes, especially in the Middle Causeway where many of the columns have a hexagonal, or six-sided, cross-section.

The Isle of Staffa (*below*), like many small islands in Scotland's Inner Hebrides, was born some 50 million years ago when lava blanketed the region after prolonged volcanic activity within the Earth's crust. As the lava slowly cooled, Staffa was left with cliffs and caves fashioned from closely-packed, hexagonal columns similar to those at the Giant's Causeway.

Staffa lies some 11km (7mi) off the west coast of the Isle of Mull and is renowned for the largest and most spectacular of its seabound grottoes – Fingal's Cave. The ceiling, walls and submarine floor of this cavern are almost entirely lined with dark grey basalt columns. When the tide is high or the sea rough, Fingal's Cave resonates with a rhythmic yet plaintive sound.

CHEDDAR GORGE

Ravine carved through England's west country

When, in 1794, the English poets Samuel Taylor Coleridge and Robert Southey visited Cheddar Gorge they were temporarily locked up in a garret on suspicion of being muggers. An unruffled Southey later stated that 'the cliffs amply repaid us'. Coleridge, inspired by the region's caves, included the lines 'Where Alph, the sacred river, ran / Through caverns measureless to man / Down to a sunless sea' in his poem *Kubla Khan*.

Cheddar Gorge cuts deep into the southern escarpment of the Mendip Hills, which stretch for about 48km (30mi) across the north of the county of Somerset in southwest England. At their highest, these limestone hills reach 315m (1,066ft). The craggy cliffs of the gorge rise more than 122m (400ft) above the winding road below. In places, bare, rugged limestone shows through the profuse vegetation, but elsewhere the cliffs are shrouded in greenery. From the hilltop, the gorge winds down for about 3.2km (2mi), twisting and turning before reaching the town of Cheddar at its base.

At the beginning of the 19th century, scientists concluded that the gorge had been created by a titanic earthquake which ripped the Mendips apart in a single colossal upheaval. By the close of the century, however, this explanation had been replaced by two others. According to one theory, continuous rainfall had percolated down through the limestone and formed an underground river which etched away, and then flowed through, myriad caves and passages. As the percolation continued the river grew larger and the rocky roof above it became thinner. Eventually, the roof collapsed, as at Verdon Gorge in the southwest of France, opening the enormous cave to the sky.

The theory that is more generally accepted points to the fact that the gorge resembles a dry river valley. During the ice ages of

Cheddar Gorge lies toward the western end of England's Mendip Hills, some 11km (7mi) northwest of Wells and 24km (15mi) southwest of Bristol (*left*). The rocky walls of the impressive cleft (*right*) soar upward to heights of 137m (450ft), or three times the height of Nelson's Column in London's Trafalgar Square. Limestone caves are situated at the gorge's lower end where one of Britain's largest underground rivers, the Cheddar Yeo, emerges from the hills.

the past million years, when all but the surface of the land was permanently frozen, a river cascaded over the bare limestone and gradually carved its way to the plain below. The river dissolved the underlying limestone, creating caves, sinkholes and passages, before it dwindled to a stream and flowed underground.

The limestone show caves

Until 1837, the spectacular gorge was Cheddar's only attraction. But in that year, when local miller George Cox was quarrying limestone beside the mill, he discovered a huge hole in the rock. Upon entering, Cox was staggered to see a magnificent cavern adorned with stalactites and stalagmites.

The entrance cavern contains stalagmites stained either red from iron oxide or blue-black from manganese. In another cave, a triple stalactite named the Peal of Bells emits a musical note when lightly tapped. At the deepest point of the cave system, which runs for 90m (295ft), lies a sinkhole of unfathomable depth. The water in this hole rises and falls throughout the year depending on the water level of the underground stream which emerges at the mouth of the gorge.

The success of Cox's cave system inspired other Cheddar residents to search for hidden natural wonders. In 1890, Richard Gough broke through into the largest and most spectacular caverns yet discovered at Cheddar, a system which penetrates 1,143m (3,750ft) into the Mendip hillside. In the entrance cave a man's skeleton was discovered during excavations in 1903. Nicknamed the Cheddar Man, this skeleton dates back to around 10,000 BC and indicates that the caves were inhabited during the early Stone Age.

Deeper still into this subterranean wonderland, where the temperature remains a constant 11°C (52°F), Gough unveiled fantastic natural formations which resembled frozen waterfalls, rivers of marble, pillared halls and cathedral vaults. Many of them were tinted green from copper carbonate, grey from lead or red from iron oxide. Gough gave grand or romantic names to the formations as he showed visitors around the caves. A delicate collection of white pillars mirrored by a still pool of water became Aladdin's Cave. A large, airy chamber, some 21m (70ft) high, was named St Paul's, after the cathedral in London.

The gorge's endangered habitat

Cheddar Gorge harbours a unique and isolated community of plants on its precipitous slopes and limestone ledges. Two of these, the Cheddar pink (*Dianthus gratianopolitanus*) and the Cheddar hawkweed (*Hieracium stenolepiforme*), are found nowhere else in Britain. The pink, which lives on rocky ledges, was a particular favourite with Victorian tourists on account of its sweet-smelling blooms. In the 1980s, it is an endangered species. The gorge is also one of the country's richest habitats of the whitebeam (*Sorbus aria*) and several related trees.

British conservationists have realized this rare plant community is in jeopardy. In the first half of the 20th century, the bare, rocky gorge provided a perfect environment for the plants; ever since, the hillsides have been invaded by rough scrub which threatens to swamp the original vegetation. This invasion came about after the deaths of millions of rabbits by myxomatosis, which meant that opportunistic bushes and young shrubs were no longer being eaten. Conservationists are hoping to turn the clock back by removing much of the scrub and so preventing Cheddar Gorge from being transformed into an ordinary wooded valley in which rare plants, such as the Cheddar pink, and butterflies have vanished.

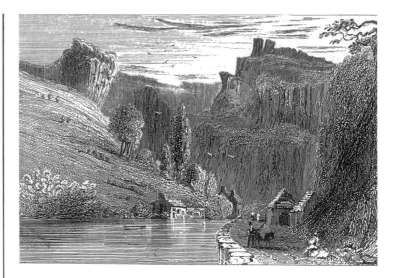

The limestone crags and pinnacles of Cheddar Gorge have been, since medieval times, the chief attraction of the area.

The discovery of the show caves in the 19th century transformed the gorge into one of England's most visited natural wonders.

The first underground formation to be discovered at Gough's show cave was a succession of giant steps that became known as the Fonts. These stalagmite basins ascend to the ceiling and were formed by the slow deposition of calcium carbonate from water.

The Speaker's Mace stands in the heart of Cox's show cave. Its name comes from its resemblance to the sceptre, or symbol of royal authority, on display in Britain's House of Commons. The corrugated stalagmite is unusual because it is wider at the top than at the base.

GLOSSARY

A

Alluvial fan: the fan-shaped deposit of loose sediment, known as alluvium, laid down by a stream or river as it emerges from a narrow gorge or valley into an open plain.

B

Barchan: a crescent-shaped sand dune with 'horns' pointing downwind. A barchan is built up by a steady prevailing wind which blows sand grains over the top and around the edges of the dune.

Basalt: hard, dark grey rock made up of fine grains and formed from solidified *lava*. The many types of basalt are common in all parts of the world.

Butte: a flat-topped hill, characteristic of southwestern USA, in which a layer of hard rock protects lower, softer layers of rock from some of the forces of erosion.

C

Calcareous: relating to rocks or soils containing a high level of the mineral calcium carbonate.

Caldera: an immense *crater* formed after a volcano explodes and collapses in on itself. The crater, which may contain the remains of a still active volcano, is surrounded by steep cliffs and often filled partially or totally by a lake.

Conglomerate: heterogeneous rock made up of rounded pebbles and sediment bonded together.

Continental drift: the gradual movement of continents across the face of the Earth over long periods of geological time. Such movement is brought about by the displacement of gigantic 'rafts', or *tectonic plates*, that make up the Earth's crust.

Crater: the funnel-shaped basin at the summit of a volcano's cone; a crater also describes the hollow created by the impact of a large *meteorite*.

Crevasse: a deep vertical crevice in the surface of a *glacier*.

D

Dolomite: a hard rock, known as magnesian limestone, composed largely of the minerals calcium carbonate and magnesium carbonate.

F

Feldspar: a widespread silica-based mineral which constitutes much of the Earth's crust.

Fjord: a long, deep, narrow inlet from the sea formed after a *glacier* has gouged out the bottom of an already-established river valley.

G

Gabbro: a hard, coarse-grained rock formed from solidified *lava*.

Geothermal: relating to the heat generated within the Earth's crust by the underlying molten rock, or *magma*.

Geyser: a hot spring which, when underground waters are periodically heated to boiling point by *geothermal* energy, is forced upward under pressure into a temporary jet of water.

Glaciation: the process of becoming covered with one or more *glaciers*; the effects of glacial action on a landscape.

Glacier: a mass of ice, formed by an immense depth of compressed snow, which gradually moves down a valley under the force of gravity.

Gneiss: a coarse-grained, *metamorphic rock* which is characterized by bands of light and dark minerals.

Graben: a valley formed by rifting, in which land is sinking between two parallel geological fault lines.

Granite: a hard, coarse-grained rock, rich in *quartz* and *feldspar*, and formed by the slow solidification of *lava*.

Gully: a long channel etched by water, especially on the side of a hill.

I

Ice age: a period of intense *glaciation* in which huge ice sheets and *glaciers* cover much of the Earth's continents, and both the level and temperature of the oceans fall.

K

Kainozoic Era: otherwise known as the Cainozoic or Cenozoic era, this comprises the last of four eras of geological time and dates from 65 million years ago to the present.

Karst: a limestone landscape riddled with caves, underground streams and passageways, which have been fashioned by slightly acidic rainwater.

L

Lagoon: a body of shallow, often tranquil, water which is isolated from the sea by a strip of land, such as a coral reef.

Lava: hot, molten rock, or *magma*, which has risen from beneath the Earth's crust and has been expelled through the vents and fissures of a volcano.

Limestone: a *sedimentary rock* composed chiefly of the mineral calcium carbonate.

Loess: fine *silt* or dust blown by winds from one region and deposited in another. A particular feature on the fringes of central Asia's arid lands, loess forms thick beds of porous, fertile soil.

M

Magma: the molten, silica-rich rock enclosed within the *mantle* below the Earth's crust except when expelled as *lava* during volcanic eruptions.

Mangrove: a tree growing beside brackish water on muddy sea shores and river banks in the tropics.

Mantle: the layer of the Earth lying between the crust and the core. A mixture of liquid and solid matter, the mantle generates convection currents which are responsible for moving the *tectonic plates* of the crust.

Mesa: a flat-topped hill.

Mesozoic Era: the third era of geological time, dating from 225 million to 65 million years ago.

Metamorphic rock: a rock, either *lava*-based or *sedimentary*, which has been transformed in character and appearance by extremes of heat, pressure or another force. Thus, heat may cause *limestone* to turn to marble, pressure may turn *shale* to slate.

Meteor: a fragment of interplanetary material, which ranges in size from a grain of sand to a massive 100 tonnes or more.

Meteorite: a *meteor* that falls through the Earth's atmosphere and strikes the ground.

Moraine: the fragments of rock and other debris generated by the erosive power of a *glacier*.

P

Palaeozoic Era: the second era of geological time, dating from 570 million to 225 million years ago.

Peat: a brown or black, fibrous, organic material composed of decaying vegetation which, because of a lack of oxygen, is prevented from breaking down completely.

Precambrian Era: the first era of geological time, dating from 4,500 million to 570 million years ago.

Pumice: a light, porous volcanic rock produced when steam and other gases bubbled through *lava* which was solidifying.

Q

Quartz: a common mineral consisting largely of *silica* and occurring in a variety of forms. The predominant constituent of most sands, quartz often contains mixtures of other minerals and ores.

Quartzite: a hard, non-porous, resistant rock derived from sandstone and composed almost entirely of *quartz*.

S

Sandstone: a porous, *sedimentary rock* composed of sand grains cemented together by such substances as *quartz* or calcium carbonate. Poorly cemented sandstones are easily eroded while strongly cemented ones are remarkably durable.

Savanna: a tropical grassland with scattered trees which experiences a wet and dry season each year, and can support grazing animals, often in large numbers.

Schist: a type of *metamorphic rock* characterized by distinct banding and derived from either a *sedimentary* or volcanic rock which has been exposed to intense heat and pressure.

Sedimentary rock: a rock deposited as a compacted bed of sediment, either rock debris, crystals or organic matter. Often laid down as sediments under water, these rocks harbour most of the world's minerals as well as fossils.

Seismic: relating to an earthquake.

Shale: a soft, fine-grained *sedimentary rock* produced from compacted or cemented mud or clay.

Silica: a common mineral known scientifically as silicon dioxide. It is the chief constituent of sand.

Silt: a deposit of fine-grained particles laid down in lakes or the water of slow-moving rivers.

Sinkhole: a depression, formed by the action of slightly acidic rainwater on *limestone* rock, through which water may enter the ground and flow along subterranean courses.

Stalactite: a column of calcium carbonate hanging, like an icicle, from the roof of a *limestone* cave.

Stalagmite: a column of calcium carbonate which grows upward from the floor of a *limestone* cave and is formed by the continual dripping of water from above.

Stratosphere: the layer of the Earth's atmosphere between 15km (9.5mi) and 50km (31mi) above the ground.

T

Tectonic plate: a structural plate of the Earth's crust which continually moves, causing earthquakes, volcanoes and *continental drift*.

Tephra: material ejected during a volcanic eruption and containing cinders, *pumice* and fine ash.

Travertine: a deposit of calcium carbonate, usually from the waters of a hot spring.

Tundra: a terrain found at high latitudes in the northern hemisphere having no trees and a permanently frozen subsoil.

V

Vent: one of a branching network of channels at the heart of a volcano through which *magma* is expelled from the *mantle* of the Earth.

Geological time has endowed every continent with a rich assortment of natural wonders, whether they be volcanoes, rivers, deserts, canyons, mountains or lakes. This gazetteer provides essential details about further natural wonders throughout the world, indicating the individual profile and setting of each. Their geographical positions are shown on the map on pages 6–7.

The gazetteer is divided into three regions: **The Americas**, including North, Central and South America; **Europe/Africa**, including western and eastern Europe; **The Near East, Far East and Australasia**, including the USSR and the Pacific Ocean.

Bryce Canyon
Utah, USA
Long. **112.3W**   Lat. **37.4N**

The centre of a national park in southern Utah, this network of canyons is a fairytale landscape of towering rock pinnacles of varying shades of red, flanked by fluted columns and sheer walls.

The canyons run for almost 32km (20mi) along the eastern side of the Paunsaugunt Plateau, which is composed of layers of sandstone, limestone and shale. When these rocks were forming on the bed of an ancient sea, they became mixed with large quantities of coloured minerals. Of these, iron compounds produce Bryce Canyon's fiery reds, copper ones create the greens and manganese compounds the deep purples. The canyons were named after local rancher Ebenezer Bryce, who described them as 'a hell of a place to lose a cow'.

Great Salt Lake
Utah, USA
Long. **113-112W**   Lat. **41.7-40.7N**

This enormous saline lake spreads for some 2,600sq.km (1,000sq.mi) to the northwest of Utah's Salt Lake City. It stands in a shallow depression 1,270m (4,167ft) above sea level, and although fed by several major rivers it has no outlet.

The lake is the remains of Lake Bonneville, which was formed after the last ice age when the glaciers melted. Bonneville once covered an area almost 20 times the size of the Great Salt Lake and reached a maximum depth of around 300m (1,000ft).

The constant loss of water through evaporation has concentrated the mineral salts brought in by the rivers so that the lake is six times saltier than the sea. Rainfall in the region varies from year to year. The lake's average depth is 4m (13ft), but in wet years this increases dramatically. The numerous islands and low-lying areas become flooded creating problems for the bison and gulls that inhabit them.

Mojave Desert
California, USA
Long. **118-116W**   Lat. **36-34.5N**

The arid lands of the Mojave Desert lie to the northeast of Los Angeles and occupy an area of around 38,850sq.km (15,000sq.mi). They are all but surrounded by mountains, notably the Sierra Nevada to the north and west. To the southeast the Mojave merges with the Colorado Desert. Its hills and gullies are composed of sedimentary rocks which contain the minerals responsible for the desert's rich colours.

The average annual rainfall in the Mojave is 12.5cm (5in), and most of the total arrives in winter. For much of the year, the arid landscape is punctuated by cacti; but when the infrequent rains do fall, they cause the Mojave to erupt in short-lived bloom.

Crater Lake
Oregon, USA
Long. **122.2W**   Lat. **42.9N**

The deep blue waters of Crater Lake lie at an altitude of 1,879m (6,164ft) in the Cascade Mountains, some 50km (31mi) northwest of Medford. They are 590m (1,940ft) deep, making Crater the second deepest lake in the Western Hemisphere after Canada's Great Slave Lake. The lake covers an area of 54sq.km (21sq.mi) and is ringed by sheer cliffs that tower to heights of 600m (1,968ft).

The lake was created some 7,000 years ago when the volcano Mount Mazama exploded, leaving behind a caldera which filled with water. The most prominent feature of the lake is Wizard Island, a volcanic peak some 237m (778ft) high, which has formed in modern times.

Crater Lake has no outlet, yet its level hardly rises or falls: rainwater falling on to its surface is almost exactly matched by the amount lost through evaporation.

Devil's Tower
Wyoming, USA
Long. **104.7W**   Lat. **44.6N**

The colossal landmark of Devil's Tower stands in an isolated region of Wyoming, some 112km (70mi) west of Deadwood. The huge monolith rises 265m (869ft) above the surrounding forest and can be seen from a distance of 160km (100mi). The flat-topped, fluted tower measures 305m (1,016ft)

wide at its base and narrows to 76m (250ft) at its peak.

The assembly of rock columns dates back some 50 million years. Then, volcanic eruptions forced a mass of molten rock upward through the floor of an ocean. As the molten rock cooled, it assumed a columnar appearance similar to that of the pillars of Giant's Causeway in Ireland. When the ocean bed was uplifted by movements in the Earth's crust, its soft, sedimentary rock was worn away, leaving behind the weather resistant columns of volcanic rock.

Blue Ridge Mountains
Eastern USA
Long. **84-77.1W**   Lat. **40.2-34.5N**

The Blue Ridge Mountains run for around 1,000km (630mi), from northern Georgia, through North Carolina and Virginia, to southern Pennsylvania. When seen from a distance the range's unbroken line of peaks appears blue, due to the scattering of light in the region's clear air.

The range is part of the Appalachian Mountains, which were thrust upward some 300 million years ago and once towered higher than the modern Himalayas. Erosion and weathering have reduced them to their present heights; in the Blue Ridge, the highest peak is Grandfather Mountain in North Carolina, which rises to 1,810m (5,938ft).

Plants, such as rhododendron, spruce and huckleberry, which thrived throughout the Blue Ridge during the last ice age, now inhabit the higher slopes where the climate remains cool.

Everglades
Florida, USA
Long. **82-80W**   Lat. **27-25N**

The Everglades are swamplands that rest on a vast plateau covering some 10,240sq.km (4,000sq.mi) of southern Florida. Their water comes from Lake Okechobee to the north which, with an area of 1,810sq.km (700sq.mi), is the third largest freshwater lake in the USA.

The humid, subtropical climate of the Everglades encourages the growth of plants and a profusion of waterfowl and other wildlife. Saw grass spreads across the shallower water; mahogany and live oaks prosper on hammocks, or low sandy islands. Where the freshwater mingles with the sea, an extensive mangrove forest—one of the largest on Earth—has evolved.

Alligators and crocodiles are the largest reptiles; birds are represented by several hundred species, such as flamingo, roseate spoonbill, egret, osprey, cormorant, ibis and wood stork. Since 1900, the bird population of the Everglades has been reduced by 90 per cent due to the decline in water.

Rabbitkettle Hotsprings
Northwest Territories, Canada
Long. 127W Lat. 62N

Rabbitkettle Hotsprings are located beside the South Nahanni River, around 256km (160mi) to the northwest of Fort Liard. Water bubbles up through a glistening, scalloped dome of calcium-rich travertine rock.

The dome, Canada's largest, stands about 27m (90ft) tall and measures 69m (228ft) across. It has been created over the last 10,000 years by mineral-rich water which, after percolating through underlying limestone and dissolving calcium carbonate, rises above ground. As the water cools from its original temperature of 21°C (70°F), the mineral precipitates out and hardens into rock at a rate of 2.5cm (1in) in every 10 years.

Cacahuamilpa Caverns
Mexico
Long. 99.6W Lat. 18.6N

The entrance to Mexico's largest caves is located in a narrow valley near Taxco, some 150km (94mi) southwest of Mexico City. Underground rivers and streams, by dissolving calcium carbonate from the surrounding limestone hills, have fashioned chambers of remarkable size and great beauty.

The largest cave at Cacahuamilpa (the name is Aztec and means 'the place where the coca grows') is 1,380m (4,530ft) long and 70m (230ft) high. Stalagmites as tall as 40m (130ft) and as broad as 20m (66ft) stand like domes on the cavern floors. Many stalagmites have been christened with names such as La Puente de los Cuerubines (the Gate of the Cherubs) and Las Fuentes (the Fountains).

Angel Falls
Venezuela
Long. 62.5W Lat. 6N

The Angel Falls, the highest waterfall in the world, were unknown to the outside world until 1935. In that year, they were discovered by US pilot Jimmy Angel, while exploring the remote Guiana Highlands in eastern Venezuela. In 1949, an expedition confirmed Jimmy Angel's claims that the falls were in fact the world's tallest.

The river on which the falls are located is the Rio Carrao, which drains the plateau of Auyan Tepuí, or Devil's Mountain, more than 640km (400mi) west of Guyana's capital, Georgetown. The main, unbroken drop measures 807m (2,648ft), while additional cataracts make up a total of 979m (3,212ft)—a figure 20 times as high as Niagara Falls. As the water of the Rio Carrao plunges over the red sandstone escarpment, it breaks up into an immense spray of water droplets, which drenches the surrounding rain forest.

Orinoco River
Venezuela/Colombia
Long. 68-61W lat. 10-3N

The remote source of the Orinoco was discovered in 1944 by aerial survey. The river was found to rise near Mount Délgado Chalbaud in the Serra Parima range of the Guiana Highlands, roughly on the border between Venezuela and Brazil. Its total length has yet to be established accurately; estimates vary between 2,410km (1,500mi) and 2,735km (1,700mi).

In its upper reaches, the Orinoco is joined to the River Amazon by a natural, navigable canal, the Casiquiare River. This channel was discovered by the German naturalist Alexander von Humboldt in 1800. The river then flows in a great curving arc through rain forest and savanna and empties into the Atlantic Ocean near Trinidad via a delta, which covers around 20,200sq.km (7,800sq.mi).

Lake Guatavita
Colombia
Long. 73.5W Lat. 5N

Lake Guatavita lies in the Colombian Andes, some 50km (31mi) northeast of the country's capital, Bogota. It is perfectly circular and surrounded by precipitous scrub-covered hills. The origins of the basin in which the lake sits have not been satisfactorily explained: the absence of lava and ash argues against a volcanic origin. Nor is there any evidence of iron or nickel that would suggest the lake occupies a crater made by a meteorite.

The lake is around 100m (330ft) deep and 400m (1,320ft) in diameter. A great notch cut into one of the slopes around the lake was made in 1580 by an expedition led by a gold-hungry Spaniard named Antonio de Sepulveda. He tried to drain the lake to reach the gold of Eldorado reputed to be lying on the lake's floor.

The quest for gold at Guatavita began after rumours that a local tribe, the Muisca, initiated their new ruler by smearing him with gold and then set him on the waters of the lake. Here, he would cast gold and emeralds into the water to appease the gods.

Cotopaxi
Ecuador
Long. 79.5W Lat. 0.7S

The snowcapped volcano of Cotopaxi rises to 5,897m (19,348ft) in the Ecuadorean Andes, some 56km (35mi) south of Quito, the country's capital, and 80km (50mi) south of the equator. Like Japan's Mount Fuji, Cotopaxi is a strato-volcano and has a symmetrical shape. Below its thickly-glaciered slopes, the mountain is made up of layers of solidified lava and ash, which have been produced by a long series of eruptions that began around 70,000 years ago.

The last major eruption from its crater, which is 610m (2,000ft) wide, was in 1877. The emission of frequent columns of smoke, plus regular minor eruptions reveal that Cotopaxi remains active. In addition, the volcano generates unusual avalanches of snow, water and mud, as the heated mountain melts the lowest layers of snow and sends thousands of tonnes of debris hurtling to the valleys below.

Atacama Desert
Chile
Long. 71-69W Lat. 28-22S

The Atacama Desert runs north-south for some 1,100km (700mi) in northern Chile, almost to the border with Peru. Beside the Pacific Ocean in the west, a mountain range rises to 1,500m (4,950ft) and is marked by sheer cliffs and ravines. The central Atacama plain stands 1,000m (3,300ft) above sea level and stretches toward the Andean foothills in the east.

The desert is reputed to be the world's driest—in some areas there has been no rain in living memory, and throughout the region there is scarcely any vegetation. Moisture-laden winds from the Amazon basin are prevented from reaching the desert by the Andes; similar winds from the Pacific Ocean are forced to drop their rain in the sea by the cold Humboldt Current that flows northward from the Antarctic.

Aconcagua
Argentina
Long. 70W Lat. 32.5S

Aconcagua is not only the highest mountain in the entire Andes range, but it is also the highest peak in the Western Hemisphere. Towering to a height of 6,960m (22,835ft), Aconcagua overlooks the border between Chile and Argentina, around 110km (70mi) northeast of Chile's capital, Santiago.

The mountain is a massive peak of volcanic rock. It stands on a foundation of extremely ancient sedimentary rock, which has been excessively deformed by upheavals in the Earth's crust. Shaped by immense glaciers, which still rest on its mountain crown, Aconcagua was first climbed by Matthias Zurbriggen in 1897. Since that year, many climbers have ascended the mountain, despite savage storms and gales of up to 250kph (156mph).

EUROPE/AFRICA

Mount Etna
Sicily, Italy
Long. **15E** *Lat.* **37.7N**

Mount Etna is Europe's highest and oldest active volcano. Its often snowcapped summit rises 3,260m (10,700ft) above the eastern coast of Sicily, some 152km (95mi) east of Palermo. Etna first erupted around 2.5 million years ago from beneath the sea; since that time, it has changed its centre of activity many times, resulting in a multitude of tiny cones and craters on the slopes of the modern mountain.

Etna's first recorded eruption, in 475BC, was noted by the ancient Greek poet, Aeschylus. Its most recent eruption took place in 1979, when ash and lava spewed forth and rolled down the mountainside. The constant production of lava by Etna means that its slopes are fertile and provide good soil for vines and other crops. Above some 2,000m (6,562ft), however, Etna's slopes are barren and black, except when covered by snow.

Lake Maggiore
Italy/Switzerland
Long. **8.5E** *Lat.* **46.2-45.7N**

Lake Maggiore is Italy's second largest lake and arguably the most beautiful lake in the Alps. It straddles the border between Italy and Switzerland, some 56km (35mi) northwest of Milan. It covers an area of 212sq.km (82sq.mi), measures 65km (40mi) in length and has an average width of 4km (2.5mi). Its maximum depth is around 370m (1,220ft).

Formed by a glacier, which retreated at the end of the last ice age some 10,000 years ago, the lake is fed by the Ticino River. The seasonal changes in the river's flow, due mainly to snow meltwater from the Alps, mean that Lake Maggiore's water level may rise and fall throughout the year by more than 1m (3.3ft).

Apart from the lake's magnificent alpine setting, its main feature is the group of four Borromean Islands near the town of Stresa.

Armand Cave
France
Long. **3.4E** *Lat.* **44.2N**

Armand Cave lies beneath the limestone hills of the Cevennes, some 80km (50mi) northwest of Montpelier in France's Languedoc region. The cave was discovered in 1897 by French pot-holer Louis Armand. After scrambling down a vertical shaft for some 75m (246ft), he entered the huge chamber which has become known as La Grande Salle (the Great Hall). This cavern measures around 100m (330ft) in length and 55m (180ft) wide.

Across the sloping floor of Armand Cave is a remarkable array of some 400 stalagmites, resembling a small forest of young conifers. The tallest measure 30m (100ft) in height. These have been formed by the slow yet persistent percolation of mineral-rich water through fissures in the roof of the limestone cave.

Camargue
France
Long. **4.2-4.7E**
Lat. **44.3-43.6N**

The Camargue is an immense marshland region that stretches for 96km (60mi) along the southern coast of France to the west of Marseilles. It is formed by the multitude of channels of the River Rhône's delta, and covers an area of 72,875 hectares (180,000 acres). The Camargue, like Las Marismas in southern Spain and the Danube Delta in Romania, provides an important crossroads for birds migrating to and from Europe. It is also a wetland sanctuary for resident birds, such as flamingo, heron, ibis, egret, marsh harrier and many species of duck.

The marshland, which consists of countless lagoons cut off from the sea by sand bars, has been formed by sediment and silt brought by the Rhône. Much of the northern Camargue has been reclaimed, while much of the southern region is covered with reeds. White horses, which were introduced by man and have become half wild, roam the Camargue in small herds, as do the black bulls destined for local bullfights.

Ben Bulbin
Ireland
Long. **8.5W** *Lat.* **54.4N**

The flat-topped bulk of Ben Bulbin rises 527m (1,730ft) above the coastal plain surrounding Sligo Bay in northwest Ireland. Legend honours it as the home of Fionn mac Cumhaill and his 3rd-century warriors; to the Irish poet W.B. Yeats it was a major source of inspiration.

The mountain's horizontal layers of sedimentary rock were laid down around 320 million years ago. Since that time, the glaciers of many ice ages, and the might of Atlantic storms, have eroded and shaped the soft, black shale that makes up Ben Bulbin's steep, corrugated walls. But the mountain's hard, upper layers, composed of thick Dartry limestone on top of thin Glencar limestone, have remained almost intact. Here, on Ben Bulbin's flat summit, lies a layer of peat covered by heath and rubble; a few arctic-alpine plants left over from the last glaciation dwell in sheltered areas.

Chesil Bank
England
Long. **2.6W** *Lat.* **50.5N**

Chesil Bank is a massive sea wall of shingle that stretches 26km (16mi) from Abbotsbury to the Isle of Portland on England's southern coast. A reef based on blue clay and covered in pebbles, Chesil Bank is in places some 11m (35ft) high and more than 135m (443ft) broad. On its sheltered, landward side lie tranquil brackish waters known as the Fleet, which are the home of Britain's largest population of mute swans.

Powerful currents and turbulent seas from the west deposit pebbles of various sizes along the beach: the smallest, which tend to be yellowish-brown in colour, remain at the western end; the pebbles gradually increase in size toward the east until at Portland, they are some 6cm (2in) in diameter and greyish in colour.

Plitvice Lakes
Yugoslavia
Long. **15.8E** *Lat.* **44.8N**

The Plitvice Lakes lie in the Dinaric Alps, some 120km (75mi) south of Zagreb in northwestern Yugoslavia. Hidden away in the forested hills, and at the heart of a national park, the series of 16 blue-green lakes has been formed at the confluence of two streams which cascade down the limestone hillside into a deep gorge. The largest lake, Kozjak, has a surface area of 85 hectares (210 acres) and a depth of 45m (148ft).

The water carries large amounts of calcium carbonate, which is forced out of solution by aquatic mosses to form a giant staircase of dams, in a process similar to that at Band-e Amir in Afghanistan. Each 'step' of the staircase may be as high as 60m (200ft) and forms the precipice for a waterfall, as the waters of one lake cascade into another. On leaving the last lake, the water falls into the sheer-sided canyon of the Korana River.

Eisriesenwelt Cave
Austria
Long. **13.1E** *Lat.* **37.5N**

The Eisriesenwelt is the world's largest, permanently ice-covered cave—its name means 'world of the ice giants'. The entrance to the Eisriesenwelt lies 1,655m (5,430ft) above sea level in the Austrian Alps, some 48km (30mi) south of Salzburg.

The cave system was discovered in 1879 and extends for 42km (26mi) into the Alps. It was etched from the limestone by weakly acidic water during the warm periods between ice ages. The cave's stalactites, stalagmites, columns, curtains and other fairytale 'sculptures' are formed not of travertine, as in many other caves, but of ice. The temperature in the cave system remains at or below freezing point throughout the year, causing dripping water to turn to ice almost as it falls.

Waddenzee
Netherlands
Long. **5-6E** *Lat.* **53.5-53N**

Before storms in the 13th century broke through the natural barrier of the West Frisian Islands on the Netherland's northwestern shoulder, the Waddenzee was a huge freshwater lake. Since that time, it has become a wetland environment, half-land and half-water, and a sanctuary for migrating birds. Estimates suggest that at the height of their annual migration more than 500,000 water fowl take refuge in the Waddenzee.

Until the 1930s, Waddenzee's waters were part of the Zuider Zee, a vast saltwater arm of the North Sea. Then, a huge dike, some 32km (20mi) long, divided the Zuider Zee into the Waddenzee and a freshwater lake, the Ijsselmeer. The construction of several polders (areas of reclaimed land) has reduced the size of the lake and reclaimed around 2,330sq.km (900sq.mi) of arable land. Protests from those who wish to see the Waddenzee maintained as a wildlife refuge have caused the cancellation of plans to reclaim land from it.

Victoria Nyanza
Uganda/Kenya/Tanzania
Long. **31.5-34.5E** *Lat.* **0.5N-2.5S**

Victoria Nyanza, also called Lake Victoria, straddles the border between three countries. Its total area of 68,600sq.km (26,500sq.mi) makes it Africa's largest lake and the second largest freshwater lake in the world after Lake Superior. Victoria Nyanza measures 336km (210mi) in length, 250km (155mi) in width and 75m (250ft) in depth. It is the main headwater reservoir for the Nile, the river which forms the lake's only outlet.

The lake occupies a depression on the Equatorial Plateau, at an altitude of 1,135m (3,725ft), between the two main branches of the Great Rift Valley. The waters of the lake are richly endowed with many species of fish. The dense human population living around the lake's deeply-indented, reed-covered shores grow crops, such as cotton, sugar, coffee and corn.

Victoria Falls
Zimbabwe/Zambia
Long. **25.9E** *Lat.* **17.9S**

The Victoria Falls are located on the Zambezi River on the border between Zimbabwe and Zambia, around 140km (87mi) upstream from Lake Kariba. The Zambezi plunges a maximum of 128m (420ft) over a precipice 1,700m (5,575ft) wide. Thus the falls are twice as wide and twice as deep as Niagara Falls.

Throughout the year, the average flow of water over the falls, and into a gorge barely 75m (246ft) wide, is 935,000 litres (205,947 gallons) per second. The local Kalolo tribe named the falls Mosi-oa-tunya, 'the smoke that thunders'. The name refers to the plume of spray that reaches 300m (1,000ft) into the air and may be seen from a distance of 40km (25mi).

Lake Chad
Chad/Cameroon/Nigeria/Niger
Long. **13-15.5E**
Lat. **14.5-12.5N**

Chad, Africa's fourth largest lake, is a body of freshwater, in contrast with most desert lakes, which are brackish. It lies on the southeastern fringes of the Sahara Desert and is located at the borders of four countries. Divided into north and south basins by a ridge known as the Great Barrier, the lake is rarely more than 7.6m (25ft) deep. In the 19th century, a depth of 285m (935ft) was recorded.

More than 2 million years ago, the ancestral Lake Chad had a surface area greater than 260,000sq.km (100,000sq.mi). In the 20th century, this area had been reduced by 90 per cent to some 16,000sq.km (6,000sq.mi). (After heavy rains, this figure may be almost doubled; in severe drought it may be all but halved.)

This reduction has been caused in part by the increasingly hot and dry climate in the centre of Africa's landmass. It has also been brought about by an uplifting of the land to the south; this altered the course of rivers that once fed the lake and diverted their flow into the Congo River basin.

Mount Kilimanjaro
Tanzania
Long. **37.4E** *Lat.* **3S**

The snow-clad peak of Mount Kilimanjaro is the highest in Africa. Located near the Kenyan border, Kilimanjaro (which in Swahili means 'mountain that glitters') lies some 340km (212mi) south of Nairobi and more than 480km (300mi) northwest of Dar es Salaam. Towering over rich wildlife plains, the huge massif is a conglomeration formed by three volcanoes which have erupted over the last two million years. Of these, Shira at 3,778m (12,395ft) is the oldest and smallest; Mawenzi stands in the centre and is 5,354m (17,564ft) high; Kibo is the youngest and tallest at 5,895m (19,340ft).

Zones of vegetation define the mantle of Kilimanjaro. Coffee and corn are cultivated on the mountain's lower slopes. Rain forest reaches up to around 3,000m (9,842ft). Grassland and moorland give way at 4,400m (14,460ft) to desert where only lichens live. Above 4,600m (15,092ft), the mountain is permanently iced over.

Ahaggar Mountains
Algeria
Long. **5-7E** *Lat.* **24-23.3N**

The lunar landscape of the Ahaggar Mountains covers some 550,000sq.km (212,000sq.mi), almost the size of the US state of Colorado. They are located in the far south of Algeria, at the heart of the Sahara Desert. Bounded to the north, east and south by the steep sandstone cliffs at the edge of the Tassili Plateau, the craggy mountains rise, at Tahat, to a height of 2,992m (9,816ft).

The Ahaggar, known also as the Hoggar Mountains, contain metamorphic rocks that are around 2,000 million years old—some of the oldest exposed rocks in Africa. Throughout the mountains, there are tall volcanic plugs; these hard cores of rock once plugged volcanic vents and are all that remain of volcanoes weathered away by erosion. The tallest of these, the Ilamen, stands 2,670m (8,760ft) above broken stones and sand on the floor of the arid landscape.

NEAR EAST/FAR EAST and AUSTRALASIA

Negev Desert
Israel
Long. 34.5-35.5E
Lat. 31.5-29.5N

A desolate region of southern Israel, the Negev covers around 13,310sq.km (5,140sq.mi), equivalent to half the country's total land area. Its name comes from the Hebrew word meaning 'dry'. The outline of the desert is shaped like an inverted triangle: its northern fringes link the Dead Sea to the fertile plains beside the Mediterranean Sea; it reaches a point, only 10km (6mi) wide, north of Eilat on the Gulf of Aqaba. Along its eastern edge runs the Wadi Arabah, a dry riverbed some 160km (100mi) long; along its western side stretches the Sinai Peninsula.

The Negev is rich in natural gas, phosphates and copper, and until the 7th century AD, when the climate turned drier, was a fertile grain-growing area. The central Negev is distinguished by crater-like depressions orientated northeast to southwest; these may be 37km (23mi) long, 8km (5mi) wide and as deep as 300m (1,000ft).

Red Sea
Middle East/Africa
Long. 32.5-43E   Lat. 30-13N

The Red Sea is part of the Great Rift Valley, which runs from Mozambique to Syria, and covers an area of 438,000sq.km (169,000sq.mi), roughly ten times the size of Switzerland. It is bordered on the African side by Egypt, Sudan and Ethiopia, and on the Middle Eastern side by Israel, Saudi Arabia and the Yemen Arab Republic. It measures some 2,100km (1,310mi) long and up to 304km (190mi) wide, and is surrounded by desert and steppes. The Red Sea receives no water from rivers.

The Red Sea's salty water, which comes solely from the Gulf of Aden through the Strait of Bab el-Mandeb, reaches a summer temperature of 29°C (85°F). Evaporation is over 200cm (80in) per year. Its central region is the deepest—its maximum is 2,134m (7,000ft)—while dangerous coral reefs line its margins. The Red Sea is so named because it contains algae of the species *Trichodesmium erythraeum* which, when they die, may transform the waters from a blue-green to a reddish colour.

Volga River
USSR
Long. 33-48E   Lat. 57.5-45.5N

The Volga, one of the world's longest rivers, flows for 3,700km (2,310mi) and is navigable almost throughout its course. Its drainage basin covers an area of 1,360,000sq.km (525,000sq.mi), which is around one third of European Russia and is inhabited by a quarter of the entire Soviet population.

From its source in the Valdai Hills northwest of Moscow, the Volga flows through many lakes and vast areas of steppes; after Volgagrad, it dips down below sea level and widens into a broad delta before flowing into the Caspian Sea. Two thirds of the river's water comes from melting snow, one third from rain and groundwater.

As Russia's chief thoroughfare, the Volga carries almost half of the USSR's freight; canals link it with the Baltic Sea, Moscow and the Black Sea, and several power stations use its flow to generate hydroelectricity.

Caspian Sea
USSR/Iran
Long. 47-52E   Lat. 47-37.3N

The Caspian Sea is the world's largest body of inland water. It covers an area of around 373,000sq.km (144,000sq.mi), equivalent to one and a half times the area of North America's Great Lakes combined. A huge salt lake whose surface is 28m (92ft) below sea level, the sea has a maximum depth in the south of 980m (3,215ft).

In the northern half of the Caspian Sea the water has an average depth of only 5m (17ft). Some 1,200km (750mi) from north to south, and an average of 320km (200mi) across, the sea receives three quarters of its water from the Volga. It is surrounded largely by mountains, and lies in a depression that was formed at least 250 million years ago. Midway along the eastern shore a gulf, known as Kara-Bogaz-Gol, has such a high level of evaporation that a layer of salts, some 2m (6.6ft) thick, lines its floor.

River Ganges
India/Bangladesh
Long. 77-91E   Lat. 33-22N

The Ganges, the most sacred river of the Hindus, rises in the Himalaya Mountains in the Indian state of Uttar Pradesh and flows 2,510km (1,567mi) southeastward across an immense undulating plain. To the north of Calcutta it splinters into many smaller rivers, such as the Hooghly, and links up with the Brahmaputra in Bangladesh to form the world's largest delta. The latter measures 400km (250mi) from north to south and 320km (200mi) from east to west.

The Ganges drainage basin covers an area of around 975,900sq.km (376,800sq.mi), roughly a quarter of the Indian subcontinent. Flowing through one of the world's most densely populated areas, the Ganges carries an enormous cargo of sediment: at an annual average of some 2.4 billion tonnes, this is more than any other river.

Western Ghats
India
Long. 73-77E   Lat. 21-8N

The Western Ghats is a mountain range fringing India's western coast. It runs for some 1,600km (1,000mi) from the mouth of the Tapti River, north of Bombay, to Cape Cormorin at the southern tip of India. The only break in this mountain wall is the Paighat Gap, around 32km (20mi) wide, in the southern part of the range. The Ghats' western edge rises steeply to average heights of between 900m (3,000ft) and 1,500m (5,000ft), and is separated from the Arabian Sea via a narrow, fertile coastal strip.

Around 60 million years ago, when India was colliding with Eurasia, a fault line developed along western India and the lava that was expelled turned to basalt. At the same time, India was lifted up in the west and tilted toward the east. As a result rivers, such as the Godavari and the Krishna, rise in the Western Ghats yet flow eastward across gradual, forested slopes and undulating plains to the Bay of Bengal.

Mekong River
Southeast Asia
Long. 96-107E   Lat. 33-10N

One of the world's longest rivers, the Mekong flows for 4,180km (2,600mi) through southeast Asia. It rises in the Tibetan Highlands, at an altitude of 5,000m (16,400ft), and turbulently descends the mountains through narrow rugged gorges. As it flows along the Thailand–Laos border, the Mekong meanders southeastward across undulating plains. In northern Kampuchea it becomes a broad waterway laden with silt. It flows through Vietnam into the South China Sea via a huge delta, one of Asia's greatest rice-growing areas.

Monsoon rains and melting snow are the river's main water inputs: in early spring the river is a trickle; by late summer it is a raging torrent. Much of the floodwater flows into Tônlé Sap, a huge lake to the north of Kampuchea's capital, Phnom Penh. This lake acts as a natural regulator of the river, absorbing excess water in summer and adding to the river in winter.

Lake Toba
Indonesia
Long. **99E** *Lat.* **2.5S**

Toba, the largest lake in Indonesia, has a surface area of 1,645sq.km (635sq.mi). The blue-green waters of this oval-shaped lake lie in a depression in the Batak Highlands in northwest Sumatra, some 160km (100mi) south of Medan.

The lake is around 90km (56mi) long and 30km (19mi) wide, and is surrounded by steep cliffs and pine-covered slopes. Formed some 60,000 years ago when a volcano exploded and collapsed, leaving behind a caldera, the lake is one of the highest, at 900m (2,953ft), and deepest, at 450m (1,476ft), on Earth. Much of the lake is filled by the island of Samosir, which was formed by volcanic eruptions and is linked to the mainland by an isthmus.

Gobi Desert
China/Mongolia
Long. **96-118E** *Lat.* **46-42N**

The Gobi, one of the world's largest deserts, covers an area of around 1,295,000sq.km (500,000sq.mi), almost the size of the US state of Alaska. It stretches east to west for about 1,610km (1,000mi) across southeast Mongolia and northern China, from the Da Hinggan Ling Mountains to the Tian Shan Mountains.

Lying on a plateau between 910m (3,000ft) and 1,520km (5,000ft) high, the Gobi is sandy in the west and composed largely of alkaline flats in the east. Much of its surface is covered with small stones known as *gobi*.

Prevailing northwesterly winds have removed most of the soil and deposited it in north-central China as loess. A small number of Bactrian camels roam wild in the Gobi, summering in the Altai Mountains to the northwest and spending the cold winters on the desert floor.

Huang He River
China
Long. **96-118E** *Lat.* **37.5-33N**

The Huang He, the second longest river in China, rises in the Kunlun Mountains in northwest Qinghai Province and flows for 4,830km (3,000mi) before emptying in a gulf of the Yellow Sea known as the Bohai Sea. The river flows eastward from the mountains before flowing north around the plateau of the Ordos Desert.

As the Huang He flows south again, it passes through the Shanxi loess region, a fertile land distinguished by the yellow silt brought from the Gobi Desert to the north and west. From this point the river carries a cargo of yellow silt, hence its former name of the Yellow River. After flowing through the Dragon Gate, a gorge 20km (12mi) long and only 15m (50ft) wide at its narrowest, the Huang He broadens dramatically and meanders eastward across the Great China Plain. Here, in China's agricultural heartland, the river bed may be raised as much as 18–21m (60–70ft) above the surrounding plain, which is prone to disastrous floods.

Mount Mayon
Philippines
Long. **123.4E** *Lat.* **13.2N**

Mount Mayon, one of the Philippines' 10 active volcanoes, is composed of the accumulated layers of lava and ash, and has a symmetrical cone to rival Japan's Mount Fuji. Rising 2,421m (7,943ft) above the Albay Gulf, Mount Mayon stands on the southeastern tip of Luzon, the Philippines' main island. Its circular base has a circumference of 130km (80mi); its crater has a diameter of 500m (1,640ft) and a depth of 100m (330ft).

Manila hemp plantations exploit the high rainfall and fertile soil of the north and western slopes; but the slopes to the south and east, riddled with gullies, support little vegetation.

The volcano's first recorded eruption was in 1616; its most devastating was in 1814 when 1,200 people died and the town of Cagsawa, 16km (10mi) to the south, was destroyed by an avalanche of lava. Since 1616, the volcano has erupted 30 times.

Mariana Trench
Pacific Ocean
Long. **142-148E** *Lat.* **23-11N**

The Mariana Trench, the lowest point on the Earth's surface, is a deep, curved scar on the floor of the western north Pacific Ocean to the east of the Mariana Islands. Measuring 2,550km (1,592mi) long, with an average width of 69km (43mi), the trench curves southeast from the small island of Iwo Jima.

The deepest point in the submarine canyon is at Challenger Deep, which lies 11,033m (36,198ft) below sea level, some 338km (210mi) southwest of the island of Guam. This point was reached on January 23, 1960, by Swiss engineer Jacques Piccard and oceanographer Lt Don Walsh of the US Navy in the bathyscape *Trieste*. Through the thick plastic portholes of this vessel, the explorers saw a flat, featureless plain which Piccard described as 'a vast emptiness beyond all comprehension'.

Beppu
Japan
Long. **131.3E** *Lat.* **33.2N**

More than 3,500 hot springs, geysers and steam vents are congregated in Beppu in northeastern Kyushu, Japan's westernmost island. In AD 867, an eruption of Mount Tsurumi in this intensely volcanic region triggered the flow of Beppu's waters, which bubble up at a rate of 45.4 million litres (10 million gallons) a day.

Boiling ponds, known as *jigokus*, emit gases as well as mineral-rich hot water. The *Chinoike jigoku* gives forth blood-red water, owing to its high level of iron oxide; the *Tatsu-maki jigoku* has a geyser which erupts every 17 minutes.

People with diverse medical complaints visit the hugely popular spa for treatment. Arthritis sufferers, for example, are buried in hot sand baths, known as *sunayus*. In addition, the source of geothermal energy is being tapped: scientists estimate that there is enough energy to produce 40 million kilowatts of power for around 1,000 years.

Matsushima Bay
Japan
Long. **141E** *Lat.* **38.3N**

Located around 320km (200mi) north of Tokyo on the Pacific coast of Honshu, Matsushima Bay is richly endowed with dozens of pine-covered islands and islets. Upheavals in the Earth's crust some 250,000 years ago pushed layers of volcanic rock above the surface of the sea.

Erosion and weathering forged a landscape of hills and valleys which, at the end of the last ice age around 10,000 years ago, was inundated by the sea as the polar ice caps melted. Thus the bay was formed and the hills became islands; the sea later fashioned rugged cliffs and crags, as well as intricate caves and tunnels. Most islands are uninhabited but have lyrical names, such as Question and Answer, or Entrance of Buddha into Paradise.

Simpson Desert
Australia
Long. **135-139E** *Lat.* **27-23S**

Long, parallel dunes of red sand characterize the Simpson Desert, which lies mainly in Northern Territory, but also part of southwest Queensland and northeast South Australia. Stretching from the saline Lake Eyre to the Macdonnell Ranges, the desert covers an area of 145,000sq.km (56,000sq.mi), roughly the size of England and Wales combined.

The desert's colour comes from iron oxide which covers each individual grain of sand. The dunes, created by the wind, are orientated northwest to southeast; they may be up to 290km (180mi) long and 30m (100ft) wide, with troughs as broad as 300m (1,000ft).

The desert is the last refuge of mammals such as the fat-tailed marsupial mouse. Shrubs and spinifex grasses are the predominant vegetation, but when the rain falls the desert blooms temporarily. The average annual rainfall is 15–20cm (6–8in), which is spread throughout the year.

NATIONAL PARKS OF THE WORLD

National parks and protected areas exist partly because people feel in danger of losing touch with nature and with themselves as a part of nature. An awareness of the need to act, urgently, to protect our natural heritage first surfaced in the United States toward the end of the last century.

There, in a country richly endowed with incomparable natural wonders, a few farsighted individuals pressed for government action. The causes for their concern were largely the rapid disfiguring of the landscape in the wake of new settlers, expanding railroads and industry, and the decimation of the indigenous population through the wholesale slaughter of their life-sustaining bison herds. On March 1, 1872, Yellowstone was officially designated 'a public park and pleasuring ground, for the benefit and enjoyment of the people'. So the national parks movement was born.

The spread of national parks
The idea soon caught on, particularly in those countries with large wilderness areas vulnerable to exploitation. Parks were subsequently established in Canada, Australia, New Zealand and South Africa – thanks largely to the influence of the British. Mexico, with El Chico, and Argentina with Nahuel Huapi and Iguaçu, were also quick to adopt the notion of national parks.

By the end of the 19th century, momentum was gathering on an international level. Even in Europe, where land was often at a premium, and where most large open tracts had for centuries been privately owned, accelerating industrial growth sparked protest. Individual initiatives to create parks and prevent the despoliation of the countryside were also taken.

The United States, pioneers of the movement, were the first to realize that the national park concept could only work in practice, and be meaningful in the long term, if professionally organized management was available to control a site and maintain its integrity. The US National Parks Service, established in 1916, continues to set standards of excellence in park administration and protection that are the envy of the world.

Setting fundamental standards
The term national park has been applied rather loosely in some countries to areas which, far from being a wilderness, support a high degree of human activity. In an effort to impose international criteria and thus enforce certain standards, the IUCN (International Union for Conservation of Nature) agreed in 1969 that to merit the designation 'national park' a site should: cover a certain minimum area (depending on density of population); be of scientific/educational/recreational interest and contain natural landscapes of great aesthetic value; be protected against human exploitation and occupation as far as possible; be open to visitors. All these stipulations to maintain the integrity of a site are to be implemented at the highest level of authority in the country.

In November 1972 the Convention for the Protection of World Cultural and Natural Heritage was set up to provide 'a framework for international cooperation in conserving the world's outstanding natural and cultural properties'. The title 'World Heritage Site' is one that is much coveted.

Since the 1960s there has been a marked increase in public awareness of the need to protect our natural heritage. However, the integrity of national parks is threatened by urban, industrial and commercial expansion, and also by sheer numbers of visitors. Thus, Great Basin – America's newest national park – may not long continue as that country's 'last true wilderness'.

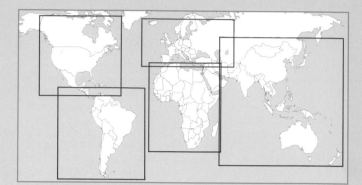

NORTH AMERICA

SOUTH AMERICA

EUROPE

- Sarek
- Rondane
- The Peak
- De Hoge Veluwe
- Brecon Beacons
- Hautes Fagnes Eifel
- Bayerischer Wald
- Swiss
- Hohe Tauern
- Vanoise
- Gran Paradiso
- Danube Delta
- Peneda Ceres
- Ordesa
- Samaria Gorge

FAR EAST and AUSTRALASIA

- Stolby Zapovednik
- Akan
- Badkhyrsky
- Fuji-Hakone-Izu
- Nanda Devi
- Corbett
- Sagarmatha
- Kaziranga
- Wen Chun Wolong
- Royal Chitawan
- Gir Lion
- Ruhunu
- Ujung Kulon
- Gunung Lorentz
- Kakadu
- Carnarvon
- Kosciusko
- Tongariro
- Cradle Mt.
- Urewara Paparoa

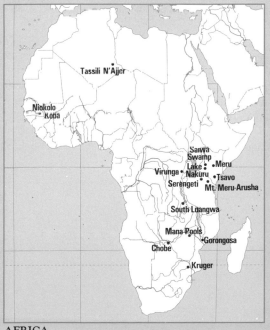

AFRICA

- Tassili N'Ajjer
- Niokolo Koba
- Saiwa Swamp
- Meru
- Lake Nakuru
- Virunga
- Tsavo
- Serengeti
- Mt. Meru-Arusha
- South Luangwa
- Mana Pools
- Gorongosa
- Chobe
- Kruger

These five maps – of North and South America, Africa, Europe and the Far East with Australasia – indicate the locations of the national parks and other reserves profiled on the following pages.

THE AMERICAS

The world's first national park was established in 1872 as the result of a campfire discussion at Yellowstone, Wyoming. Those who first mooted the idea could hardly have imagined that by 1986 Yellowstone would be visited by almost 2·5 million people a year. Today the United States has more parks, reserves and protected areas than any country in the world. It also manages them with greater efficiency, imagination and sensitivity than any other.

Canada, too, with its vast wilderness areas, magnificent mountains and extensive forests, offers a number of national parks on as majestic a scale as many in the United States. Quick to follow the trail blazed by Yellowstone, Canada created the splendid parks at Banff (1885) and Waterton Lakes (1895). These now form part of a chain of national parks in the Selkirk and Rocky Mountains. Waterton Lakes, together with Glacier National Park over the border in Montana, became the world's first international peace park.

Parks of South and Central America

Though the countries of South America compare unfavourably in economic terms with the nations of the northern continent, they are endowed with an equal variety of natural wonders. And they have not failed to recognize them, though current world concern is strongly focused on the relentless decimation of many South American rain forests.

At the general assembly of the IUCN at San José, Costa Rica, in February 1988, it was announced that a peace park would be created, stretching the length of the common border between Costa Rica and Nicaragua. Formerly occupied by guerilla camps or by Nicaraguan soldiers, the area is to be returned to the indigenous people, who are charged with its custody. Organized into cooperatives, they are to protect the park and its endangered species. Logging rights to 3,120sq.km (1,200sq.mi) of virgin rain forest are to be cancelled. The peace park project is to receive financial support from Scandinavian governments.

REDWOOD National Park
California, USA
Established 1968
Area 442sq.km (171sq.mi)

This park in northern California has one of the finest remaining virgin stands of coastal redwood. Of the three kinds of redwood, two grow only in California, the third in central China. The coastal redwood, found along the Pacific coast from Big Sur to the Oregon border, differs from the sierra redwood (giant sequoia) that grows on the eastern slopes of the Sierra Nevada. The former grows taller, but the latter grows wider and can live longer. (The Chinese dawn redwood is smaller and, unlike its cousins, deciduous.)

The redwood's extraordinary regenerative powers, and its amazing resistance to fire and disease, account in part for its longevity and size. Trees over 90m (300ft) are common. One of their number is the world's tallest tree: it stands 112m (367ft) high and is found in this park. Not only do redwood attain a great age – some alive now were living at the time of Christ – they may also achieve enormous girth.

YOSEMITE National Park
California, USA
Established 1890
Area 3,079sq.km (1,189sq.mi)

Yosemite, a spectacular area in California's Sierra Nevada, is a land of soaring peaks, thundering waterfalls, towering trees, rushing rivers and quiet meadows. This is a park of superlatives – Yosemite Falls, at 739m (2,425ft), is the highest in the United States and among the highest in the world. El Capitan, symbol of the park, is the world's most massive chunk of solid granite. The park contains the world's greatest collection of granite domes.

There are three groves of the magnificent giant sequoias – the largest living things in the world. Old Grizzly, at 2,700 years old, is the fifth largest tree in the world. Yosemite's fauna include the black bear, bobcat and grey fox.

GREAT BASIN National Park
Nevada, USA
Established 1986
Area 491,400sq.km (189,000sq.mi)

This is the first new national park to be created in the United States since 1971, and the first ever in Nevada. It stretches from Utah to California and from Oregon to Arizona, and is one of the country's last real wildernesses. Much of this vast landscape is covered with small mountain ranges that are too dry and rocky to support much life. But in some areas desert quickly gives way to alpine meadows as, at higher altitudes, the land springs into life.

Wheeler Peak, which rises to 3,981m (13,063ft), is the park's highest mountain, and near its summit is the southernmost permanent ice field in the USA. This rugged and remote area is nonetheless rich in animal and plant life; the shaggy bristlecone pines – the most long-lived organisms on Earth – are found here. The oldest pine of all, aged 4,900 years, was cut down in 1964 – in the interests of science. Lehman Caves national monument and Lexington Arch, a huge limestone formation at the south end of the park, are among the park's star attractions.

CARLSBAD CAVERNS National Park
New Mexico, USA
Established 1930
Area 189sq.km (73sq.mi)

Every evening, from April to September, millions of bats emerge from the entrance to Carlsbad Caverns, the largest known cave system in the world. It was the bats who first drew attention to the caves, and their evening exodus has now become a tourist attraction.

These ancient caverns in the Guadalupe Mountains boast the world's greatest variety of limestone formations. Beside splendid stalactites, stalagmites, soda straws and columns – such as the Temple of the Sun, dripping with tiers of flowstone – there are geological rarities, such as cave pearls, lily pads, and epsomite needles.

Most impressive of all is the Big Room, the largest known underground chamber in the

United States. This covers an area of up to 5.6 hectares (14 acres) and is tall enough to accommodate a 30-storey building. Some 309m (1,013ft) below the surface, at the lowest known point in the caverns, is the Lake of Clouds. Here, the water is always fresh because, although no water flows in or out of it, the lake supports no life forms at all.

CANYONLANDS National Park
Utah, USA
Established 1964
Area 1,365sq.km (524sq.mi)

Canyonlands is the strangest, most spectacular of landscapes. From the Grand View Point at the Island in the Sky, between the Green and Colorado rivers, the eye spans a vast, inhospitable array of bare, flat-topped mesas slashed by deep canyons which converge (at the Confluence) in the centre of the park.

Among the park's abundance of bizarre features is the Land of the Standing Rocks. Also known as the Maze, this is a land of tortured labyrinths watched over by multicoloured rocky sentinels. The tallest of the park's many arches is David's Arch, which stands 110m (360ft) high. There are huge monoliths, reminiscent of the standing stones of England's Stonehenge, and incredible stone spires soaring to more than 90m (300ft). As well as all the natural sculptures, superb ancient Indian paintings adorn the walls of Horseshoe Canyon.

ZION National Park
Utah, USA
Established 1919
Area 593sq.km (229sq.mi)

The work of nature as both sculptor and painter is constantly in evidence at Zion. The effects of iron on sandstone produce pink, red and brown washes on vast stone murals. The effects of wind, weather and water create arresting patterns, textures and contours in the rocks.

The Virgin River is the master sculptor in this land of towering mesas and deep canyons. One spectacular effect of natural erosion can be seen in the Narrows at the head of the canyon; in places these are only 6m (20ft) wide and more than 300m (1,000ft) deep.

Among Zion's many wonders is the world's largest free-standing natural arch – the Kolob Arch, spanning almost 95m (310ft). One of the park's most extraordinary landmarks is the checkerboard mesa, on which horizontal lines of ancient dunes are intersected by weathered fissures to produce a regular, almost geometric, pattern.

YELLOWSTONE National Park
Wyoming/Montana/Idaho, USA
Established 1872
Area 8,984sq.km (3,469sq.mi)

Attempts to describe the world's first national park rapidly run out of superlatives. This uniquely beautiful and varied landscape simply overwhelms the visitor with its wonders.

Nowhere else on Earth is the planet's molten interior as close to the surface as it is here. Yellowstone has the world's greatest collection of geysers (about 300 in all), of which the most famous is Old Faithful, adopted as the park's symbol. It is unique among geysers because its pattern of eruption has hardly changed for 100 years, and this dependability has earned it its name. After a brief prelude of sporadic spurts, Old Faithful will suddenly blast up to 60m (200ft) heavenward. It may hurl some 38,590 litres (8,500 gallons) of water into the air during the two to five minutes it is in action.

The drama of Yellowstone is truly audio-visual, with constant plopping, bubbling, hissing and splashing accompanying the countless thermal manifestations of a huge 'subterranean pressure cooker'. Competing for attention with the giant geysers are the travertine balconies and terraces of Mammoth Hot Springs in the park's northwest corner; equally impressive are the 27 layers of petrified forests exposed in the rockface at Specimen Ridge.

WOOD BUFFALO National Park
Alberta, Canada
Established 1922
Area 44,800sq.km (17,300sq.mi)

This is Canada's largest national park, and the most northerly area of prairie marsh in North America. The Peace-Athabasca delta is also one of the largest freshwater deltas in the world. The park was originally designated a protected area not least to safeguard the future of the great whooping crane, an endangered species, and has now become that bird's only known nesting ground.

Wood Buffalo, however, is perhaps best known as the sanctuary for the largest free-roaming herd of bison in the world; it numbers between 14,000 and 16,000. Black and grizzly bears, wolf, caribou, wolverine and moose are among the park's larger mammals. Small mammals include the porcupine and beaver.

NAHANNI National Park
Northwest Territories, Canada
Established 1976
Area 4,784sq.km (1,840sq.mi)

This park, the first natural area in the world to be officially designated a World Heritage Site (in 1979), includes a number of outstanding geological features. Nahanni is 'a pure and limitless wilderness', rarely penetrated by white men until recently – the fate of the few who did venture into the interior in search of gold is recorded in such ominous place names as Deadman Valley and Headless Creek. Since it is accessible only by boat and floatplane, even today the park gets relatively few visitors – though road building programmes outside its boundaries are likely to increase the numbers.

The park possesses some of the most striking karst terrain in the western hemisphere. The most famous manifestations are at Rabbitkettle Hot Springs. Here, a tufa mound, 27m (90ft) high and 69m (228ft) across, is scalloped with ornate terraces built up from minerals dissolved in the 21°C (70°F)-spring water. Other notable karst features are seemingly bottomless sinkholes, plunging canyons and vast numbers of caves – many still unexplored.

The park's other star attractions are the Virginia Falls, described by one explorer as 'pure detonation'; the strange limestone pillars known as 'hoodoos'; and the Devil's Kitchen, where whistling winds have created weird sculptures in the sand.

KOOTENAY National Park
British Columbia, Canada
Established 1920
Area 1,390sq.km (537sq.mi)

Highway 93 cuts through the middle of Kootenay, which extends 8km (5mi) either side of it for a distance of 112km (70mi). The two are intimately connected, for the park was primarily established to protect the landscape on each side of the road.

Lush green valleys, sheer red sandstone cliffs, canyons, waterfalls, lakes scattered with icebergs, and numerous hot springs are among the park's many outstanding natural wonders.

Among the star attractions is Marble Canyon: 60m (200ft) deep, the gorge has walls layered with marble. At Radium Hot Springs, 2,270,000 litres (500,000 gallons) of water gush forth each day. At a temperature of 46°C (114°F), these waters are among the hottest of all the springs in the Canadian Rockies, and have long been prized for their healing qualities.

BANFF National Park
Alberta, Canada
Established 1885
Area 6,641sq.km (2,564sq.mi)

Hot springs are an outstanding feature of many areas of the Rocky Mountains, and it was around those at Banff that the first of Canada's national parks took shape. In 1885, 26sq.km (10sq.mi) of hot springs were set aside 'for sanitary advantage to the public'.

Banff's alpine setting offers a wide range of scenic wonders: flower-bedecked meadows in forested valleys, beautiful lakes, mighty rivers and glaciers, and a number of impressive peaks. The highest peak is Mount Assiniboine, at 3,618m (11,870ft). The turquoise waters of Lake Louise lie amid green forests and soaring peaks with the Victoria Glacier in the background.

GALAPAGOS National Park
Ecuador
Established 1934
Area 6,912sq.km (2,658sq.mi)

'The natural history of these islands is eminently curious and well deserves attention,' observed Charles Darwin, ship's naturalist aboard HMS *Beagle* in 1835. The isolated volcanic archipelago where Darwin encountered the unique creatures that prompted his theory of evolution by natural selection is today a national park.

There are six main islands, a dozen smaller ones and some 40 islets; each is the peak of a massive volcanic cone, some of which are still active. Vegetation, for the most part, is sparse and the islands' main attraction is the unique fauna. Here, reptiles rule, for no land mammals reached the islands.

The absence of large predators means the islands' creatures have little instinctive fear of humans. Visitors simply walk around the nesting boobies, frigate birds and albatrosses. Huge land iguanas, marine iguanas (the only sea-going lizards in the world) and sea lions are similarly unperturbed. Giant tortoises are much reduced in number and rarely seen, since they inhabit the more inaccessible areas of the archipelago.

NAHUEL HUAPI National Park
Argentina
Established 1903
Area 7,850sq.km (1,274sq.mi)

This mountain region in the Andes offers a rich diversity of natural wonders: lakes, rivers, glaciers, waterfalls, forests, and snow-clad peaks. Vast stands of ancient trees contrast with open, flowering prairies. From December to April the park is resplendent with wild flowers. On the lake shores are foxgloves, fuchsias and lupins. Daisies, lilies and primroses carpet the fields.

The park's most outstanding natural feature is the lake after which it is named. This covers an area of 531sq.km (204sq.mi) and in places is as much as 300m (984ft) deep. Mount Tronador, the highest peak in northern Patagonia at 3,554m (11,660ft), dominates the park.

AFRICA

British and French colonial territories were the first to adopt the national parks concept pioneered in the United States. First to come into being was the Sabie Game Reserve in South Africa, founded in 1898 and redesignated the Kruger National Park in 1926. The original philosophy behind the creation of the parks was, as in the United States, to protect large wilderness areas and to preserve their natural beauty. This protection did not extend to wildlife, which often fell prey to big game hunters. Although hunting is illegal in most African reserves today, enforcing the law is usually difficult.

Parks were maintained as such after many countries obtained their independence, because they represented an important source of revenue as well as a point of prestige. Kenya, for example, earns more from tourism than from any other source, and in the decade from 1965–1975 witnessed a staggering increase in visitor numbers from 11,000 to 1.5 million.

Protecting animals

Animals are the main attraction in the African parks – and steps have been taken to ensure that they can be observed in safety. Within the parks certain areas have been developed – with safari lodges and artificial watering holes to attract game.

However, the creation of special wildlife reserves, no matter how large, is a double-edged sword. While they have literally proved a life saver to many endangered species threatened with extinction the reserves are also creating problems. In certain areas of Africa – and indeed in Europe – park wardens are faced with the task of culling animals in order to obviate problems arising from huge population increases. The defoliation of trees by African elephants is an example. In Africa the expansion of human populations on their boundaries is also threatening the parks' integrity.

TASSILI N'AJJER National Park
Algeria
Established 1972
Area 3,000sq.km (1,158sq.mi)

This national park, which is also a World Heritage Site, may be extended to a massive 70,000sq.km (27,027sq.mi) in area. Extraordinary 'rock forests' sculpted from eroded sandstone are the park's most outstanding geophysical feature. Equally impressive, and certainly of immense significance both historically and geologically, are the ancient rock paintings and engravings found in this area of the Sahara Desert.

Drawings of hippopotamuses are evidence of a time in prehistory when the region was much wetter than it is today, and could support such water-loving creatures. Buffalo, rhino, giraffe and elephant – all known to have been extinct in this region for several thousand years – are also depicted. Cattle herders shown in later paintings may have been partly responsible, through overgrazing, for the ultimate desertification of the region.

CHOBE National Park
Botswana
Established 1961
Area 10,830sq.km (4,181sq.mi)

The vast, underpopulated, generally arid country of Botswana possesses a number of magnificent wildlife reserves, including the famous Central Kalahari Game Reserve. Chobe, which extends along the Chobe, Ngwezumba and Savuti river valleys, offers a rich diversity of landscape. As well as swamp and marsh habitats, there are forests, grasslands and deserts. The Chobe River is a favourite haunt of hippos, and lions prowl the Savuti marshlands. Other predators found here include the hyena, cheetah and leopard. There are also a number of large mammals: elephant, buffalo, zebra, giraffe, and some 18 antelope species.

GORONGOSA National Park
Mozambique
Established 1920
Area 3,770sq.km (1,456sq.mi)

Four different and quite distinct habitats are found in Gorongosa, which lies between a range of mountains and a plateau. Areas of marshland, which are completely inundated in the rainy season from November until April, contrast with scrub, subtropical jungle and open grassland.

The park's wildlife includes many large African mammals: hippo, elephant, black rhino, leopard, buffalo, zebra, antelope and monkey, as well as two species of crocodile. The African roller, crowned crane and spoonbill are among the 340 bird species that have been recorded in the park. Among Gorongosa's star attractions are the lions who have set up home in a group of abandoned houses.

NIOKOLO-KOBA National Park
Senegal
Established 1962
Area 8,130sq.km (3,139sq.mi)

This, the largest national park in Senegal, was until 1986 the only national park of note in the whole of West Africa. Watered by three rivers, chiefly the Gambia River, it is a huge reserve which does not revert to desert in the dry season.

Ponds, lakes and gorges all contribute to the beauty of the landscape. Tall grasses, including some superb bamboos, clothe this land of vast savannas, which provides a sanctuary for an enormously varied range of wildlife. There are a number of endangered species which exist here in healthy populations: Derby's eland, leopard, wild dog, elephant and bubal hartebeest.

Impressive statistics record 70 mammal species, 35 reptile species and birds in their thousands, with more than 325 species recorded in the park.

KRUGER National Park
South Africa
Established 1898
Area 19,485sq.km (7,523sq.mi)

The Kruger has been known under its present name since 1926. Before that date the park was the Sabie Game Reserve, the first protected area in South Africa. Still the biggest area of protected land in that country, it is a long narrow strip, running north-south. Because it is entirely fenced in, except where rivers flow into neighbouring Mozambique, it severely limits the natural migratory habits of land-dwellers.

Kruger is recognized internationally as possessing one of the finest wildlife areas in the world. A strict management programme has been introduced to control the wildlife within the park's boundaries, which involves some 'culling'. And since animals are unable to leave the park in search of water, wardens must ensure adequate artificial water sources in the dry season.

The park provides a refuge for the endangered black and white rhinos. Other large mammals include the elephant, hippopotamus, numerous antelope species, red jackal, giraffe, lion, cheetah, hyena and leopard. The ostrich is among the park's 400 recorded bird species.

MT MERU-ARUSHA National Park
Tanzania
Established 1962
Area 137sq.km (53sq.mi)

Mount Meru, which rises to 4,450m (14,600ft), is the fourth highest peak in Africa. It was a late addition to the original area of this park which can be broadly divided into three areas: woodland and bush north of Ngurdoto, the Momela lakes and the Meru crater itself. Generally rugged terrain is alleviated by lush grasslands surrounding the lakes – this is part of an unusual pattern of watersheds that ensued when Meru's caldera was formed.

One of the park's outstanding features is the Ngurdoto crater. About 2.5km (1.5mi) across, it is surrounded by forest. Described as 'Africa in miniature', Ngurdoto is reminiscent of a small-scale Ngorongoro Crater. From various observation points on the rim, the park's large and diverse wildlife population can be viewed.

The woodland areas provide a refuge for rhino, buffalo, hyena, giraffe, elephant, and many other large mammals. The best place to watch them is at Momela lakes – their watering place – where there are magnificent views of Mount Kenya and Mount Kilimanjaro.

SERENGETI National Park
Tanzania
Established 1959
Area 14,763sq.km (5,700sq.mi)

Isolated hills and kopjes (small rocky outcrops) provide some relief among the expanses of grassy steppes which clothe much of the Serengeti. This huge park is famed for supporting the largest population of plains animals in the world. As well as 2,500 elephants, 200 black rhinos have found a refuge here, along with a number of top predators – hyena, leopard and cheetah. The Serengeti also has the greatest concentration of lions in Africa.

Every year the park affords a most spectacular eye-catching phenomenon: the migration of vast numbers of antelope and their close relations. Representing about 24 species, they include about 2 million wildebeest, 900,000 Thomson's gazelle, as well as about 300,000 zebra. The migration starts with the dry season, as the animals make their way from the eastern Serengeti across the park to the river pastures along its western corridor.

MANA POOLS National Park
Zimbabwe
Established 1963
Area 2,210sq km (850sq.mi)

One of southern Africa's few remaining wilderness areas, Mana Pools occupies a woodland area along the Zambesi River. This broad expanse of woodland, which is scattered with hidden pools, lies in the river's flood plain and is surrounded by towering escarpments.

When the surrounding countryside is dry, the Mana Pools area remains lush and moist, and is an irresistible magnet to wildlife. Animals come here from as much as 100km (62mi) away, and are particularly numerous during the dry season. Among them are several endangered species – rare black rhino, leopard and elephant – as well as cheetah and wild dog. Numerous bird species have also been recorded at the pools.

SOUTH LUANGWA National Park
Zambia
Established 1938
Area 9,100sq.km (3,500sq.mi)

Of the three national parks in the Luangwa Valley, this is the most accessible, as well as being the largest in Zambia. It stretches along the west bank of the Luangwa River, where grassy areas alternate with woodland, and mountains tower on each side. The park boasts a rich variety of waterfowl along the river and beside the lakes. Most notable are the fish eagles, and the blue carmine bee-eaters. Crested cranes, saddle-bill storks and wood ibis are also found in the river valley.

The black rhinos, for which the area was once famed, have been poached almost to extinction. However the park does possess a unique giraffe subspecies, known as Thornicroft's, as well as many zebra, kudu, waterbuck and elephant. This is the best place in Zambia to see hippos out of the water.

VIRUNGA National Park
Zaire
Established 1925
Area 8,090sq.km (3,124sq.mi)

Virunga is contiguous with Uganda's Ruwenzori and Rwanda's Volcanoes National Park, and offers a wide range of different habitats. Mount Ruwenzori, crowned by permanent snows, is the tallest of the park's eight volcanic peaks, two of which are still active.

Such diverse terrain as marshlands, lava plains and savanna support a correspondingly varied flora, often presenting astonishing contrasts; on the Ruwenzori massif, Afro-alpine flora take over at higher levels from equatorial forest. In addition, it is claimed that no other park in Africa offers such a diversity of fauna; outstanding is the mountain gorilla, found in the bamboo forests.

LAKE NAKURU National Park
Kenya
Established 1961
Area 572sq.km (220sq.mi)

Lake Nakuru, one of the most beautiful of Kenya's smaller national parks, is famed for its fabulous flocks of flamingoes. They congregate to feast on the algae which flourish on their own droppings, creating a swathe of pink around the water's edge.

On occasions in the past, there have been as many as 2 million lesser flamingoes in Nakuru. In the late 1970s, however, they started to move out to other lakes in the region. The reason is uncertain, but it is thought that a rise in the lake's salinity and an increase in the number of pelicans (attracted when the waters were stocked with tilapia fish) drove them away. In the 1980s, they started to return in impressive numbers. Their presence is, however, no longer 'guaranteed'. In all, there are now many other bird species in the park; 400 or so have been recorded, many in the woodland surrounding the lake.

MERU National Park
Kenya
Established 1967
Area 880sq.km (339sq.mi)

Meru's special quality, quite apart from its captivating landscape, arises from the fact that it is one of the least frequented, and therefore the least disturbed, of Kenya's national parks.

Permanent streams in the area ensure a lush vegetation, with swamps, tall grass, forests and impressive palms. It was in Meru that Elsa, the famous lioness featured in Joy Adamson's story *Born Free*, was returned to the wild. Turtles abound in the Rojewero, the park's largest river; and elephant, buffalo, zebra and giraffe are also common.

The park's star attraction is the rare white rhino – so precious that it receives round the clock protection to preserve it from poachers. Extinct in Kenya since prehistoric times, the rhino has been re-introduced from a game reserve in South Africa. Meru is the only national park in Kenya where they can be seen.

SAIWA SWAMP National Park
Kenya
Established 1974
Area 192 hectares (474 acres)

Saiwa, the smallest national park in Kenya, was established in order to protect an endangered species of semi-aquatic antelope called the *sitatunga* (pronounced 'statunga').

The *sitatunga* is a reddish-brown creature with large ears. Its other notable features are its hooves, which are splayed and elongated. They are rarely seen, however, since the antelope spends most of its time neck-deep in the swamp. *Sitatunga* are also found in west and central Africa, and possibly round Lake Victoria.

TSAVO National Parks
Kenya
Established 1948
Area 20,821sq.km (8,039sq.mi)

Tsavo East and West together make up one of the largest national parks in the world. Of the two, Tsavo West, with well-watered scarp and volcanic landscapes, is the more frequented. Tsavo East is more arid, with vast plains of dry brush; at least two thirds of its vast area is closed to the public in order to protect its herds of elephant and rhino from poachers.

Poaching has wrought a terrible toll in both the parks at Tsavo; the rare black rhino, which numbered between 6,000 and 9,000 in 1969, had been reduced to a mere 100 by 1981.

Elephants are especially numerous in Tsavo, but many other large mammals can also be seen, particularly around the game lodges. These animals include gazelle, zebra, buffalo and oryx. Tsavo is of interest to ornithologists because it lies on an avian 'migration corridor'.

The park's principal attraction is Mzima Springs, which gush forth 2,724 million litres (600 million gallons) of water a day. The outstanding purity of the water is due to lava – most spectacularly evident in the black outcrops of the petrified Shetani flow that comes from the Shetani Mountain.

EUROPE

Geography and density of population mean that the only national parks in Europe that could be defined as 'wilderness' are found either in Scandinavia or in remote, less accessible mountain regions, such as the Alps or the Pyrenees.

Elsewhere, as in Great Britain, the term 'national park' has been adopted for areas of great natural beauty which are also inhabited, and to some extent exploited, by man. Terminology varies: in Germany, for example, such parks would be called Nature Parks.

Competition for land is intense, and these special areas may, in some instances, serve primarily as recreational areas. There are, in addition, fundamental conflicts of interest between conservationists and developers. The Lake District in England was denied in 1988 the coveted 'World Heritage' status because of inappropriate development within its boundaries.

Early moves toward preservation

Private initiative lies behind many of the several thousand parks and reserves scattered throughout Europe. Many, indeed, were first protected in the 19th century as reserves for the nobility. Italy's Gran Paradiso national park was set aside as a royal hunting domain before the world's first national park was established at Yellowstone.

The English National Trust began acquiring land as early as 1895 and the Dutch conservation group Vereiniging tot Behoud van Naturmonumenten was founded in 1905.

In Germany the Verein Naturschutzpark, founded in the early years of this century, has promoted the idea of Nature Parks – areas set aside primarily for human recreational activities, but in which the character of the landscape should be preserved as far as possible. Some such parks have specially protected areas within their boundaries.

Some of the most enlightened thinking, at government level, has come from Sweden and Switzerland – both countries with land to spare. Sweden was the first country in Europe to follow the lead of the USA when it set up four national parks in 1909.

HOHE TAUERN National Park
Austria
Established 1976
Area 2,589sq.km (1,000sq.mi)

This spectacularly beautiful alpine region of peaks, valleys and waterfalls was first proposed as a national park in 1923. Besides the Grossglockner, the highest peak in Austria at 3,797m (12,457ft), the Hohe Tauern contains one of the largest glaciers in the eastern Alps and the eighth highest waterfall in the world – the Krimml Falls. This, the most impressive of many waterfalls in the region, tumbles 400m (1,300ft) into the valley below. The highest point in Europe that is accessible by car is also here: the Edelweiss-spitze, at 2,505km (8,218ft).

Coniferous forests of spruce, pine and larch provide a home for mammals that include red deer, chamois, hare and marmot. Many species of rare bird, including the pygmy owl, golden eagle, snow finch and lammergeier come to breed in the park area.

HAUTES FAGNES EIFEL Natural Park
Belgium
Established 1973
Area 500sq.km (192sq.mi)

Hautes Fagnes Eifel is the largest of the Belgian state reserves, and situated in one of Europe's oldest forests in the Ardennes. On the border with West Germany, it is contiguous with the German-Luxembourg Nature Park. The picturesque Our Valley occupies most of the southern part of the park. Small fields and hedges of beech in the central region contrast with the vast upland forests of spruce and beech.

The Hautes Fagnes Nature Reserve is an area of marshy moorland plateau. Its peat bogs host a number of unusual marshland plants including bog asphodel and the insect-eating common sundew. Resident fauna include wild boar and wild cat as well as red and roe deer. Many interesting bird species breed in the reserve, including ring ouzels, Tengelman's owl and the great grey shrike.

SAMARIA GORGE National Park
Crete, Greece
Established 1953
Area 8sq.km (3sq.mi)

Described as one of the Seven Wonders of Europe, this spectacular gorge with its precipitous escarpments won the Council of Europe's European Diploma in 1979. From the south side of the plain of Omalos, the deep gorge descends 16km (10mi) to the sea.

The gorge is only 3.5m (11ft) wide at its narrowest point, where sheer red and grey rock walls soar upward 600m (1,968ft) on either side. The park is home to the Cretan wild goat, and has 14 endemic species of flower. The *Paeonia clusii*, with its large white blooms, it renowned as the most beautiful.

VANOISE National Park
France
Established 1963
Area 528sq.km (204sq.mi)

Lakes, glaciers and glacial features, such as moraines, are punctuated by several impressive peaks in this interalpine region bordered by the Arc and Isère valleys. The Vanoise is contiguous with the Italian Gran Paradiso National Park. The Pointe de la Grande Casse is the park's highest peak, at 3,861m (12,667ft).

Notable flora include alpine groundsel and alpine asphodel. Marmots, mountain hares, weasels and foxes are common smaller mammals; larger ones include the ibex and the chamois. The park hosts a variety of interesting bird life including the golden eagle, citril finch, ptarmigan and rock partridge.

As with other French national parks, this one is zoned; within the park itself are two nature reserves – Col d'Iseran and Tignes – and a surrounding buffer zone, all of them governed by conservation rules other than those of the park.

DE HOGE VELUWE National Park
Netherlands
Established 1930
Area 46sq.km (18sq.mi)

The Deelerwoud nature reserve forms a bridge between the De Hoge Veluwe and the Veluwezoom parks, both of which are privately controlled. The drifting sands of the De Hoge Veluwe, which formed a desert 100 years ago, have now been stabilized by tree planting.

Heathlands, inland dunes and grasslands support an interesting variety of flora. As well as the Scots pine, which is in danger of overtaking much of the heathland, oak, juniper and beech take their place alongside many rare plants, including chickweed wintergreen and dwarf vipergrass. Red deer and mouflon have been introduced to the area; native mammals include foxes, badgers, polecats, pinemarten, wild boars and roe deer.

GRAN PARADISO National Park
Italy
Established 1922
Area 700sq.km (270sq.mi)

An area of extraordinary natural beauty lying in the mountain ranges of northwest Italy, this park is contiguous with the French Vanoise National Park. Alpine peaks up to 4,061m (13,323ft) high contrast with wooded valleys and picturesque meadows. In June, the latter are studded with a profusion of alpine flowers, which attract a variety of butterflies.

Without the existence of the Gran Paradiso, the ibex would have been doomed to extinction. Now both it and the chamois exist there in such large numbers that measures are being taken to control them. At night, the ibex, which normally live at around 3,000m (9,800ft), descend to the lower reaches of the mountains and return in the early morning.

Most other large mammals, including the wolf and brown bear, were killed off by hunters before the park was established. Today, small mammals include foxes, mountain hares and brown hares. The park shelters more than 80 bird species, including the golden eagle and the eagle owl, the ptarmigan, white grouse and snow finch.

PENEDA CERES National Park
Portugal
Established 1970
Area 600sq.km (230sq.mi)

In an area long celebrated for the splendour of its scenery, this horseshoe-shaped park offers peaks and valleys, granite escarpments and mountain torrents. The mountains are modest in size – the highest is Mount Nerosa at 1,545m (5,069ft). The park's high annual rainfall guarantees a rich variety of flora: its 18 endemic species of flowering plant include the rare *Iris boissieri*. The park supports a rare species of lizard, Schreiber's green lizard, which is only found in this area of Portugal and northwest Spain.

ORDESA National Park
Spain
Established 1918
Area 157sq.km (60sq.mi)

Ordesa is contiguous with the French Pyrenees National Park and is one of the two oldest national parks in Spain. Considered in conjunction with its French neighbour, it is Europe's largest protected area.

An extraordinary variety of landscapes – waterfalls, forests, cliffs, gorges and glaciers supports a rich variety of endemic flora and fauna. The Pyrenean saxifrage and aquilegia are two flower species peculiar to the region. Small mammals include the Pyrenean desman, a strange mole-like creature with a long tail and distinctive muzzle. In the French park, the small population of brown bears is protected.

SWISS National Park
Switzerland
Established 1914
Area 169sq.km (65sq.mi)

The Swiss National Park is the model for the parks of other European countries. It is situated in a high mountain region in the Engadine valley, adjacent to the Italian national park at Stelio. Soaring peaks, valleys, streams and forests combine to make this one of the most beautiful, as well as the best managed, conservation areas in Europe.

Within this natural magnificence, the park supports a rich variety of wildlife: marmots and chamois are common, and the ibex, which was extinct here in the 19th century, flourishes again. Reindeer, which had disappeared from the park 75

years ago, have been reintroduced and are now so prolific they are overgrazing. Among the park's unusual bird species are the black woodpecker, capercaillie, eagle owl and golden eagle. Alpine meadows shimmer in early summer with dazzling displays of rock roses, gentians and orchids.

DANUBE DELTA Nature Reserve
Romania
Established 1962
Area 400sq.km (153sq.mi)

Vast numbers of migratory waterfowl from Asia and northern Europe overwinter in the Danube Delta, since the climate here is markedly milder than in the rest of Europe. The region's astonishingly varied habitats support a correspondingly diverse flora and fauna. Rivers, swamps, and lakes, as well as floating reedbeds, dunes and areas of permanently dry land, play host to geese, ducks and waders. Other inhabitants include red-breasted geese; it is estimated that most of the world population of this bird winters in the Danube Delta.

BRECON BEACONS National Park
Wales
Established 1957
Area 1,344sq.km (519sq.mi)

From the Black Mountain in the west to the Black Mountains in the east, four tilted masses of red sandstone and brown stone form some of the most attractive upland areas in Wales. Beautiful lakes, many associated with local legend, nestle in the lofty cwms, or glaciated hollows, below north-facing scarps. One such is the supposedly 'bottomless' Llyn Cwm-llwch below Pen-y-Fan, the highest peak in the region at 886m (2,907ft). Its flat summit and steep scarp slope are typical of the area. The southern part of the park is characterized by deep wooded gorges, craggy ridges and numerous caves.

Three national nature reserves within the park, as well as a number of special sites of scientific interest, have been set aside. The rich variety of animal and plant life includes pole cats and badgers, pike and eels that

grow to great size in Llangorse Lake, and a diversity of arctic-alpine plants on some north-facing hills.

THE PEAK National Park
England
Established 1951
Area 1,404sq.km (542sq.mi)

The Peak district, the first area in Great Britain to be designated a national park, provides a region of outstanding natural beauty in close proximity to a number of major industrial areas.

Situated at the southernmost point of the Pennine Chain, the park possesses some of the most beautiful hill country in England. It is broadly divided into a northern, gritstone, Dark Peak area and a southern, limestone, White Peak area. Lonely moorlands, with precipitous scarps, wide valleys, numerous rivers and waterfalls, characterize the north of the park. Wooded dales, steep gorges, extensive caves and white crags distinguish the south.

BAYERISCHER WALD National Park
West Germany
Established 1969
Area 120sq.km (46sq.mi)

One of Germany's two officially designated national parks (the other is also in the Bavarian Alps at Berchtesgaden) the Bayerischer Wald park lies on the border with Czechoslovakia. In the northwest corner of the park, which is almost entirely forested, a spectacular nature reserve surrounds the Rachelsee lake. The small reserve at Neuschonau provides a refuge for those large mammals that formerly roamed wild in the area: bison, lynx, bear and wolf.

At higher elevations, spruce is the dominant tree species. Mixed woodland predominates at lower altitudes. Here the ground vegetation is more profuse and includes the rare Hungarian gentian, a plant not otherwise found outside the Alps. A notable feature of the wetter valleys are *filze* (giant moss cushions), some as tall as 5m (16ft). The alpine snow thistle, the mountain buttercup and the sundew grow along the banks of the park's numerous streams and rivers.

RONDANE National Park
Norway
Established 1962
Area 572sq.km (221sq.mi)

Rondane, Norway's first national park, has 10 peaks rising to more than 2,000m (6,526ft), as well as many typical glacial features – canyons, cirques, moraines and precipitous slopes.

Apart from a small forested area, the park is sparsely covered with alpine vegetation. As well as 300 wild reindeer, it supports wolverine and elk. Otters and mink may be found along the watercourses. Red-necked phalarope may be seen along the southern lakes, while dotterels and golden plovers frequent the higher moorland areas.

SAREK National Park
Sweden
Established 1909
Area 1,940sq.km (749sq.mi)

Sarek is an awesomely beautiful mountain region which is so remote and inaccessible that it has remained largely undisturbed. With 4 other reserves, it comprises the largest wilderness region in Europe – covering some 8,430sq.km (3,240sq.mi) and known as Swedish Lapland.

The park contains more than 90 peaks over 1,800m (5,900ft) high, numerous spectacular gorges, almost 100 glaciers, as well as waterfalls, bogs and forests. Plants are typical of those that will tolerate arctic conditions: in the summer, alpine zones are notably bedecked with the white blossoms of mountain avens. Wolves are virtually extinct in this region, but lynx, wolverine and bear are occasionally seen. The arctic fox and the elk, being less shy, are more often sighted. Golden eagles hunt in the mountains and long-tailed skuas nest in the high tundra.

FAR EAST AND USSR

Parks in India and the Far East were generally initiated by colonial governments, largely for the protection of wildlife. The British and the French introduced the concept into the various territories they occupied, and the Americans introduced it in the Philippines.

The Chinese are showing an increasing interest in conservation issues and are establishing a national park at Xishuangbanna in Yunnan Province, with backing from the World Wide Fund for Nature. A number of reserves exist for the protection of the giant panda.

The Japanese adopted the national parks idea with enthusiasm, and their islands are richly endowed with them. Density of population and limited space mean that – except perhaps on the northern island of Hokkaido – there is no such thing as a wilderness area. Indeed, most Japanese national parks are primarily recreational areas.

National parks in the Soviet Union

The Russians, under the Tsar, also embraced the idea of conservation with enthusiasm, and set up a number of national nature reserves at a time of considerable political upheaval. The main purpose of these protected areas was to provide a refuge for the European bison. Though the area around Lake Baikal has been officially designated a national park, with vast tracts of wilderness and a low-density population, the USSR has been able to concentrate on conservation without necessarily having to create specially protected areas.

Concern for the environment has been evident since the earliest days of Soviet power, but it was not until the 1950s and 1960s that it became widespread. Natural landscapes have increasingly received legislative protection with the establishment of many nature reserves and, since 1970, national parks.

WEN CHUN WOLONG Nature Reserve
Sichuan Province, China
Established 1969
Area 2,000sq.km (772sq.mi)

This reserve, which has been designated a World Heritage Biosphere Reserve, lies on the boundary of the country's subtropical and temperate zones. Like many other reserves in China, the site, which presents a wide variety of forest and alpine habitats, provides a sanctuary for the giant panda – the animal adopted as the symbol of the World Wide Fund for Nature.

Fossil studies have shown that these animals existed 2 million years ago. The fact that they are the only mammals of this class not to have become extinct thus makes them 'living fossils'. One of the most specialized large animals in the world, the giant panda is found only in areas where there is enough bamboo to sustain it. Other endangered species that find protection here are the takin (a type of antelope) and the golden monkey.

CORBETT National Park
Uttar Pradesh, India
Established 1935
Area 520sq.km (201sq.mi)

This park, named after the famous tiger hunter, Jim Corbett, is in a superb location near the foothills of the Himalayas, in the valley of the Ramganga River. Some spectacular gorges dissect an area of open plains.

Besides its most famous denizens, the tigers, the park supports an astounding variety of fauna. Indian salmon, freshwater catfish, mugger and crocodile flourish in the watery habitats, while elephant, leopard, hyena, various species of deer, langur monkey and wild boar are found in healthy numbers on dry land.

GIR LION Sanctuary/National Park
Gujarat, India
Established 1965
Area 1,412sq.km (545sq.mi)

The national park covers 140sq.km (53sq.mi) of the total sanctuary area, which is situated on the Kathiawar Peninsula. An open hilly region supporting a wide variety of Indian fauna, it is famed as the last remaining wild habitat of the Asiatic lion, whose numbers are now dangerously reduced. Other predators found within the area include the leopard and hyena. There are also numerous deer species, wild pigs, sloth bears and a variety of birds.

KAZIRANGA National Park
Assam, India
Established 1908
Area 430sq.km (166sq.mi)

Kaziranga can boast one of conservation's great success stories: within its protective boundaries, the Great Indian one-horned rhinoceros has been saved from extinction. Once they roamed the entire Ganges Basin, but were slaughtered until their population was reduced to a mere 12 animals in 1908. The recovery of the species has been so well sustained that by the late 1980s the population numbered some 1,000 beasts. The park thus qualifies as the best remaining habitat of this rare creature.

The rhinoceros horn is prized by the Chinese who believe it to have medicinal powers. It is also much sought after for fashioning dagger hilts in the Yemen. The jungle myna birds, which commonly roost on the backs of these massive creatures, perform a grooming function and also sound the alert when danger approaches.

One reason for the rhino's success story is the annual flooding of the Brahmaputra River, which prevents human occupation of the flat swampland where the park lies.

The park also provides a vitally important refuge for the water buffalo (Kaziranga is the best remaining habitat for this creature). Swamp deer, tiger and elephant are among the other endangered species to flourish within its confines.

NANDA DEVI National Park
Uttar Pradesh, India
Established 1982
Area 630sq.km (243sq.mi)

This virtually inaccessible basin, tucked away in the fastnesses of the Garal Himalayas, must have seemed like paradise to the explorers who first encountered it in 1934. It was declared a game sanctuary as early as 1939. Soaring peaks surround the basin's lush pastures. Here, animals that had experienced little contact with humans, and thus knew no fear, quietly grazed. Among the sanctuary's notable inhabitants are the snow leopard, goral, Himalayan tahr, musk deer and blue sheep.

AKAN National Park
Hokkaido, Japan
Established 1934
Area 8,700sq.km (3,346sq.mi)

Described as Japan's 'last frontier', the island of Hokkaido is largely a wilderness of moors, lakes, virgin forest and spectacular volcanoes – two of which are still active: Mount Akan and Mount Atosanupuri.

The park, which lies within a subarctic zone, contains a number of spectacular lakes. One of these, Lake Ashu, is one of the clearest lakes in the world. It is possible to gaze as much as 40m (131ft) into its crystal depths.

The lake is also noted as the home of a unique freshwater plant – the Marino weed. A number of Ainu, the aboriginal people of Japan, live within the park area.

FUJI-HAKONE-IZU National Park
Honshu, Japan
Established 1931
Area 1,232sq.km (476sq.mi)

If sheer numbers of visitors are any guide to popularity, then this park, with 15 million a year, must be one of the most popular in the world. The park is dominated by Mount Fuji, which rises directly from the sea, and soars in splendid symmetry over vast plains. This, one of the world's most famous mountains, is an important feature of Japanese religious, social and artistic life.

The superlative beauty of the region is reflected not only in the majesty and tranquillity of the mighty mountain, but in the spectacular signs of geothermal activity. As well as countless hot springs, lava cliffs and cascades, Izu Peninsula boasts seven active volcanic islands. Scenic lakes and forests add to the outstanding natural beauty; forests of azalea, cherry, pine and fir cloak the lower slopes of Mount Fuji. The park supports a wide variety of birds, while its mammals include the Japanese dormouse.

ROYAL CHITAWAN National Park
Nepal
Established 1973
Area 550sq.km (211sq.mi)

In southern Nepal, in the outer foothills of the Himalayas, lies a region called the *terai*. This is an area of riverine forests, swamps, grasslands, lakes and duns (valleys) drained by sluggish rivers. The park is located in the Chitawan dun, drained by the Rapti River, whose annual flooding has a marked effect on the surrounding vegetation.

An outstanding feature of the park is the marsh crocodile population. It is also a refuge for the rare Great Indian one-horned rhinoceros. In 1973, when the park was established, there were only about 200 of these creatures left in Nepal. Since the imposition of heavy penalties for poaching, however, their population has increased, and by 1982 had reached 375.

The leopard, sloth bear and gaur are found here. The park also provides one of the best tiger sanctuaries on the subcontinent, and is the only tiger population to have been studied at length.

SAGARMATHA National Park
Nepal
Established 1976
Area 1,240sq.km (476sq.mi)

The idea of a national park 'at the top of the world' was first mooted in 1973 by one of Everest's conquerors, Sir Edmund Hillary. With the help of several international conservation groups, who volunteered both funds and staff, and with particular assistance from the New Zealand aid programme, the park was officially opened in 1976. Its declared aim was to 'safeguard a site of major significance not only to Nepal but to the world'. After serious initial doubts, and much local opposition, the park has begun to win the acceptance of the indigenous Sherpas.

To the Sherpas, Sagarmatha (the Nepalese name for Mount Everest) is 'the goddess of the Universe'. Everest is the highest mountain in the world, at 8,848m (29,028ft).

Among the area's notable fauna are the rarely seen snow leopard, the tahr (a gravity-defying, goatlike animal), musk deer and the yak. Most controversial of all is the yeti (abominable snowman) which is said to inhabit the region. Some of the 120 bird species recorded here include the golden eagle, the Himalayan griffon vulture, and Nepal's national bird, the Impeyan pheasant.

RUHUNU National Park
Sri Lanka
Established 1938
Area 1,090sq.km (421sq.mi)

Located in the southeast corner of the island, some 296km (185mi) from Colombo, the country's capital, Ruhunu is Sri Lanka's premier wildlife sanctuary. The park, which is more popularly known as Yala, is renowned for its elephants, leopards and bears, each species having their own place of assembly. Star attractions are the sloth bears and numerous peacocks, as well as the island's endemic species, the golden palm civet and the rusty-spotted cat.

Enormous flocks of birds are drawn to the lakes and lagoons scattered throughout the park, which has areas of scrub jungle, outcrops of rock and flat plains. Peafowl and quail frequent the plains while in the forests dwell orioles, barbets, hornbills and flycatchers.

UJUNG KULON National Park
Java, Indonesia
Established 1921
Area 786sq.km (302sq.mi)

The remnants of Krakatoa, the volcano which erupted so devastatingly in 1883, provide some of the more awe-inspiring sights in this region of low volcanic hills and plateaus, scattered with sand dunes and lagoons. A narrow isthmus joins the peninsula on which the park is situated with the western end of mainland Java.

Because of this isolation, the wildlife of the region has always enjoyed a high degree of natural protection, and its populations have flourished. The park now contains the most unusual variety of animals in the whole of Java. It is also a refuge for the rare and endangered Javan rhinoceros: about 60 live in Ujung Kulon, and they are believed to be extinct elsewhere in the world. Other threatened species offered protection on the peninsula are the Javan leaf monkey, Javan gibbon, leopard, wild ox (known as banteng) and wild dog.

GUNUNG LORENTZ Nature Reserve
Irian Jaya, Indonesia
Established 1978
Area 20,000sq.km (7,690sq.mi)

The Gunung Lorentz reserve stretches from tropical sandy beaches to the snow-clad summit of Mount Jaya, which, at 5,039m (16,532ft), is the highest mountain in southeast Asia. The enormous range of vegetation supports a number of New Guinea's endemic species, as well as a wide variety of marsupials. Birdlife is prolific and exotic; birds of paradise share the reserve with cassowaries, bower birds, parrots and cockatoos.

BADKHYZSKY Nature Reserve
Turkmen SSR, USSR
Established 1941
Area 1,330sq.km (514sq.mi)

This park is a land of startling contrasts, where jet-black mountains soar beside a brilliant white salt lake. Surrounding the lake are strange, spectacularly sculpted rock formations.

When this reserve in the southern part of Turkmenistan was established in 1941, its wild ass population was on the verge of extinction. Since then, this animal species has made a remarkable recovery and is represented by a healthy population. Other large mammals include wild goats and gazelles, wild pigs, caracals, honey badgers, wolves and hyenas.

A unique plant, the ferula, grows here. Reaching its full height in the space of six weeks, it then withers and remains dormant for the next six or more years, whereupon it suddenly springs back into life, blooms, sheds its seeds and dies.

STOLBY Zapovednik
Krasnoyarsk Kray, USSR
Established 1925
Area 472sq.km (182sq.mi)

Located in the east of the Sayan Mountain range, between the Mana and Yenisey rivers, the Stolby Zapovednik (reservation) has a taiga landscape containing spruce, larch and pine forests, as well as rhododendron and birch woodlands. The reservation's trees provide refuge for typical middle Siberian taiga animals, such as the musk deer, bear, lynx, Siberian stag and numerous small mammals. The star attraction of the reservation is the collection of unusual rock pillars, known as *stolby*. These strange sculptures are outcrops of granite and syenite (a sedimentary rock containing various silica-based minerals), which reach heights of 100m (330ft).

AUSTRALASIA

Following the American lead, the British were quick to appreciate the significance of the national park concept and to apply it to many of their overseas territories. Thus national parks were established in Australia and New Zealand soon after the one at Yellowstone: the Royal National Park in Australia was founded in 1878 and Tongariro in New Zealand in 1887, the latter mainly at the instigation of the indigenous Maori people. Since then the number of parks in Australia, the world's smallest continent, has proliferated. They encompass a wide variety of ecosystems – from the lush humidity of the tropics to snowcapped icy peaks.

New Zealand's national parks

Fjordland, Mount Cook and Westland are three of New Zealand's national parks which have been declared World Heritage Sites. It was proposed in the late 1980s that the area from Okarito in the middle of South Island's northern coast to Waitutu in the south be declared a South West New Zealand World Heritage Site.

The parks undoubtedly encompass some of the country's most spectacular landscapes. They are all, however, suffering the depredations of animals such as cats, introduced by early settlers. Since there were no native predators, the newcomers encountered no natural enemies to combat their population numbers. They have now proliferated to the extent that they pose a serious menace to local bird life as well as to the vegetative cover of the islands. Difficult terrain sometimes makes it almost impossible to implement effective control programmes.

In 1887, Chief Te Heuheu Tukino IV presented the people of New Zealand with their first national park in the form of his people's sacred peaks. By 1987, the country had a total of 12 national and 3 maritime parks. To celebrate the 100-year link with the past, New Zealand's prime minister, David Lange, presented Te Heuheu's great grandson with a framed copy of the original deed of gift.

CRADLE MOUNTAIN – LAKE ST CLAIR National Park

Tasmania
Established 1940
Area 1,319sq.km (509sq.mi)

Along with Franklin-Gordon Wild Rivers National Park, Cradle Mountain forms a World Heritage Biosphere Reserve known as the Southwest National Park. This is one of the few remaining temperate wilderness areas in Tasmania. The island's highest mountains, as well as numerous streams and lakes (including Lake St Clair), contrast with open plains and savannas, forests of eucalyptus and pine.

The park's wide range of vegetation supports a variety of wildlife. Kangaroo, wallaby, platypus, Tasmanian devil, wildcat, bandicoot, opossum and wombat all find sanctuary within its confines. It is here, so it is thought, that the Tasmanian wolf – if indeed it still survives – is likely to be rediscovered.

KAKADU National Park

Northern Territory, Australia
Established 1979
Area 13,000sq.km (5,019sq.mi)

This vast area of floodplains, punctuated with massive sandstone plateaus and escarpments, is one of the biggest national parks in the world. At the Top End of the Northern Territory, some 248km (155mi) east of Darwin, Kakadu is an aboriginal homeland; more than 1,000 archaeological sites have been found here, among them evidence of the earliest known human settlement on the continent.

A number of endangered species have been provided with sanctuary in Kakadu's rivers and forests. These include the estuarine crocodile, hooded parrot, and chestnut-quilled pigeon. One quarter of the continent's fish species are located here; most notable is a primitive form of the archer fish, whose only other recorded existence is in New Guinea. One third of the country's bird species have been recorded in the park.

KOSCIUSKO National Park

New South Wales, Australia
Established 1944
Area 6,297sq.km (2,431sq.mi)

Australia's highest mountain, Mount Kosciusko (2,228m/7,310ft), dominates this region on the western face of the Great Dividing Range. This is the largest national park in New South Wales and it contains all the state's snowfields.

In the Australian version of the Swiss Alps, there are innumerable signs of glacial activity: moraines, cirques, lakes, vast snowfields, massive plateaus and spectacular peaks. Koalas, kangaroos, wallabies and wombats all have a home in the park, as do the duck-billed platypus and the spiny anteater, although these last two are rarely sighted.

CARNARVON National Park

Queensland, Australia
Established 1938
Area 269sq.km (104sq.mi)

Carnarvon Creek gorge is the most outstanding feature of this park. Stretching 32km (20mi) on the eastern face of the Great Dividing Range, its sheer sandstone walls soar 180m (590ft) high. In some of the numerous side gorges are caves containing unspoilt examples of aboriginal art.

Massive tree ferns – some growing as tall as 12m (40ft) – spotted eucalyptus, and an impressive variety of wild orchids are among the park's most notable flora. Animals that find a refuge here are the rare brush-tailed rock wallaby, duck-billed platypus, koala and grey kangaroo.

PAPAROA National Park

South Island, New Zealand
Established 1987
Area 280sq.km (108sq.mi)

Paparoa, the newest national park in New Zealand, was opened to commemorate the centenary of the country's national parks. It offers a rich variety of landscapes and habitats – ranging from the wild and rugged headlands along the coast, to the caves, forests and mountains of the interior. The unique splendour of the region, however, resides in its karst landscape – a limestone terrain modelled by weakly acidic rainwater. Complex and beautiful, this territory is the best of its kind in the country.

While intricate weavings of cliffs and gorges make for an outstanding natural wonderland above ground, there are still more marvels beneath the surface. The most spectacular limestone formations so far discovered are those in the Metro Caves on the Nile River.

TONGARIRO National Park

North Island, New Zealand
Established 1887
Area 800sq.km (307sq.mi)

This was New Zealand's first national park, and the fourth in the world. Chief Te Heuheu Tukino IV presented it to the government on behalf of the Tuwahare people, asking only that the park be kept *tapu*, or sacred, and protected.

The Tuwahare offered their lands for protection in order to safeguard their ancestral mountains – the volcanic peaks of Tongariro, Ruapehu and Ngauruhoe in the centre of the island. These peaks dominate a surrounding landscape of varied vegetation. Contrasting with the stark, 'lunar' landscape around the peaks (which are still active), and with the arid areas of scrub, are lush alpine meadows, forests and waterfalls. On Tongariro's northern slopes at Ketetahi are found numerous hot springs.

UREWARA National Park

North Island, New Zealand
Established 1954
Area 1,995sq.km (770sq.mi)

In Urewara time has stood still. Nothing, it seems, has changed since the time of the dinosaurs. Mist-shrouded primeval forests grow rich and luxuriant, with tall ancient trees in the valleys, hardwoods and southern beeches on the more exposed slopes.

Amid the rapids and dramatic waterfalls of the Aniwaniwa stream, the rare blue duck flourishes. Lake Waikaremoana, into which the stream flows, is arguably the most beautiful lake in the country. This land has been inhabited by indigenous people – the Tuhoe – for at least 1,000 years. These are the 'children of the mist' and the park is steeped in Maori legend.

BIBLIOGRAPHY

Andriamirado, S. (1978) *Madagascar Today*. Editions Jeune Afrique, Paris.

Ayensu, E.S. (1980) *Jungles*. Jonathan Cape, London.

Baillie, K., Salmon, T. and Sanger, A. (1985) *The Rough Guide to France*. Routledge and Kegan Paul, London.

Banks, M. (1975) *Greenland*. David & Charles, Newton Abbot.

Barrington, N. and Stanton, W.I. (1977) *Mendip: the Complete Caves and a View of the Hills*. Cheddar Valley Press, Somerset.

Bolles, E.B. (1979) *Fodor's Animal Parks of Africa*. Hodder & Stoughton, London; (1979) McKay, New York.

Burgis, M.J. and Morris, P. (1987) *The Natural History of Lakes*. Cambridge University Press, Cambridge.

Carrington, R. (1967) *Great National Parks*. Weidenfeld and Nicolson, London.

Conrad, J. (1973) *Heart of Darkness*. Penguin, Harmondsworth; (1984) Penguin Bks., Inc., New York.

Cooper, D. (1970) *Skye*. Routledge and Kegan Paul, London; (1983) Methuen, Inc., New York.

Crump, D.J. (Ed.) (1983) *Nature's World of Wonders*. The National Geographic Society, Washington, D.C.

de Jongh, B. (1979) *The Companion Guide to Mainland Greece*. William Collins Sons & Co. Ltd, London.

Deschamps, M. (1987) *Travels in Provence*. Phoenix Publishing Association, Bushey; (1988) David & Charles, Vermont.

Diallo, S. (1977) *Zaire Today*. Editions Jeune Afrique, Paris.

Douglas, N. (1986) *Fountains in the Sand*. Oxford University Press, Oxford.

Drew, D. (1985) *Karst Processes and Landforms*. Macmillan Educational, London.

Duffey, E. (1982) *National Parks and Reserves of Europe*. Macdonald, London.

Durrell, L. (1986) *State of the Ark*. The Bodley Head Ltd, London; (1986) Doubleday & Company Inc., New York.

Frater, A. (Ed.) (1984) *Great Rivers of the World*. Hodder & Stoughton, London; (1984) Little, Boston.

Freely, J. (1986) *The Companion Guide to Turkey*. William Collins Sons & Co. Ltd, London.

Fridrikson, S. (1975) *Surtsey*. Butterworth & Co., London.

Garside, E. (1981) *China Companion*. Andre Deutsch, London; FS&G, New York.

Heuvelmans, B. (1965) *On the Track of Unknown Animals*. Granada Publishing Ltd, London.

Hunt, J. (1953) *The Ascent of Everest*. Hodder & Stoughton, London.

Huntford, R. (1979) *The Last Place on Earth*. Hodder & Stoughton, London; (1985) Atheneum, New York.

Huxley, A. (Ed.) (1962) *Standard Encyclopedia of the World's Mountains*. Weidenfeld and Nicolson (Educational) Ltd, London.

Innes, H. (1964) *Scandinavia*. Time-Life International, Netherlands.

Inernational Union for Conservation of Nature and Natural Resources (1982) *The World's Greatest Natural Areas*. IUCN, Switzerland.

King, T. (1981) *In the Shadow of the Giants*. Tantivy Press, London; (1981) A.S. Barnes & Company, Inc., New York.

Learmouth, A. and Learmouth, N. (1973) *Encyclopedia of Australia*. Frederick Warne and Co. Ltd, London.

McGregor, C. (1974) *The Great Barrier Reef*. Time-Life International (Nederland) B.V.

Mitsikosta, T. (1984) *Meteora*. The Holy Convent of St Stephen, Meteóra, Greece.

Moorehead, A. (1960) *The White Nile*. Hamish Hamilton, London; (1983) Random, New York.

Moorehead, A. (1962) *The Blue Nile*. Hamish Hamilton, London; (1983) Random, New York.

Morris, P. and Farr, C. (1985) *The Rough Guide to Tunisia*. Routledge and Kegan Paul, London; (1985) Methuen, Inc., New York.

Morton, H.V. (1984) *In Search of Scotland*. Methuen, London; (1984) Dodd, New York.

Newby, E. (1985) *A Book of Travellers' Tales*. William Collins Sons & Co. Ltd, London.

Oey, E. (Ed.) *Indonesia*. APA Productions (HK) Ltd, Hong Kong.

Pennington, P. (1979) *The Great Explorers*. Aldus Books Ltd, London.

Pilkington, R. (1961) *Small Boat through Sweden*. Ian Henry Publications, Hornchurch.

Pringle, L. (1985) *Rivers and Lakes*. Time-Life Books Inc., Amsterdam.

Ross, A. (1986) *Pagan Celts*. B.T. Batsford Ltd, London; (1986) B&N Imports, New York.

Sears, D.W. (1978) *The Nature and Origin of Meteorites*. Adam Hilger Ltd, Bristol.

Severin, T. (1973) *The African Adventure*. Hamish Hamilton Ltd, London.

Smith, A. (1978) *Wilderness*. Allen and Unwin, London; (1978) W.H. Smith & Son Ltd, New York.

Sterling, T. (1973) *The Amazon*. Time-Life International (Nederland) B.V.

Strahler, A.N. and Strahler, A.H. (1987) *Physical Geography*. John Wiley, Chichester and New York.

Sugden, D. (1982) *Arctic and Antarctic*. Basil Blackwell, Oxford; (1982) B&N Imports, New York.

Suttles, S.A. and Suttles Graham, B. (1986) *Fielding's Africa South of the Sahara*. William Morrow & Company, Inc., New York.

Swire, O. (1961) *Skye*. Blackie and Son Ltd, Glasgow.

United Nations (1971) *List of National Parks and Equivalent Reserves*. Hayez, Belgium.

Wallace, R. (1972) *The Grand Canyon*. Time-Life International (Nederland) B.V.

Waltham, A.C. (1974) *Caves*. Macmillan, London.

Weare, G. (1986) *Trekking in the Indian Himalaya*. Lonely Planet Publications, Victoria and Oakland, California.

Wernick, R. (1979) *The Vikings*. Time-Life Books Inc., Amsterdam; (1979) Silver, Massachussetts.

Westwood, J. (1987) *The Atlas of Mysterious Places*. Weidenfeld and Nicolson, London and New York.

Whitfield, P. (Ed.) (1984) *Illustrated Animal Encyclopedia*. Longman Group Limited, Harlow; (1984) Macmillan Publishing Company, New York.

Yapp, P. (Ed.) (1983) *The Travellers' Dictionary of Quotation*. Routledge and Kegan Paul, London; (1985) Methuen, Inc., New York.

Youde, P. (1982) *China*. B.T. Batsford Ltd, London.

INDEX

Bold numbers indicate main essays; italicized numbers indicate captions to illustrations.

INDEX

ACKNOWLEDGEMENTS

The publishers would like to thank the following, from whom they received invaluable assistance in the compilation of this book:

Donald Binney
Simon Blacker
Hendrina Ellis
Ken Grange, New Zealand
    Oceanographic Institute
International Union for
    Conservation of Nature and
    Natural Resources, Cambridge,
    UK
Shelley Turner
Jazz Wilson

All maps by Oxford
Cartographers, Oxford, UK

All illustrations by Vana Haggerty,
except illustrations on p.94 and
p.135 by Tony Graham.

PICTURE CREDITS

l=left; *r*=right; *t*=top; *c*=centre;
b=bottom

8/9 The Image Bank; 10/11 M. Huet/ Hoa-Qui; 12/13 Zefa Picture Library; 14*l* Antikvarisk Topografiska Arkivet/ Robert Harding Picture Library; 14*r* British Museum/Robert Harding Picture Library; 15 Johan Berge/ Norwegian National Tourist Office; 16/ 17 Rosine Mazin/Agence Top; 18/19 Roger-Viollet; 20/21 Le Reverend/ Explorer; 22 Hutchison Library; 23*c&b* Collection Devaux/Explorer; 23*t* John Cleare/Mountain Camera; 24/25 Ch. Errath/Explorer; 26/27 Peter Carmichael/Aspect Picture Library; 28/ 29 Tore Hagman/Naturfotograferna; 30/31 Tore Hagman/Naturfotograferna; 31 Leif Öster/AB Göta Kanalbolag; 32/33 W. Rozbros/Explorer; 34*b* Leonberg Storto/Berchtesgadener Land; 34*t* Ernst Baumann/Berchtesgadener Land; 35 GDT-Silvestris/ NHPA; 36/37 Archivio Consorzio Frasassi; 38/39 Ezio Quiresi; 40/41 Krafft/Explorer; 42/43 Robert Harding Picture Library; 43*b* Krafft/Explorer; 43*t* John G. Ross/Susan Griggs Agency; 44/45 Carol Jopp/Robert Harding Picture Library; 46/47 Anthony Bannister/ NHPA; 48/49 Shostal Associates; 50/51 Ian Redmond/Planet Earth Pictures; 51 Tony Morrison; 52/53 J. Ph. Charbonnier/Agence Top; 54 David Beatty/ Susan Griggs Agency; 55*b* Rob Cousins/Susan Griggs Agency; 55*t* Antoinette Jaunet/Aspect Picture Library; 56/57 K. Benser/Zefa Picture Library; 58 J. Allan Cash; 58/59 Travel Photo International; 60/61 Richard Packwood/Oxford Scientific Films; 62 Richard Packwood/Oxford Scientific Films; 62/63 John Cleare/Mountain Camera; 63 Mary Evans Picture Library;

64/65 Tor Eigeland/Susan Griggs Agency; 66/67 C. Skrein/M. Epp/Zefa Picture Library; 67*b* J. Allan Cash; 67*t* The Mansell Collection; 68/69 E. Streichan/Shostal Associates; 70*b* Hutchison Library; 70*t* Spectrum Colour Library; 71 Spectrum Colour Library; 72/74 Peter Carmichael/ Aspect Picture Library; 75*b* Peter Carmichael/Aspect Picture Library; 75*t* J. Allan Cash; 76/77 Michael Freeman; 78/79 R. I. M. Campbell/Bruce Coleman; 79 Brian Seed/Aspect Picture Library; 80/81 Olivier Langrand/ Bruce Coleman; 82 Christian Zuber/ Bruce Coleman; 83 Agence Nature/ NHPA; 84/85 Dr M. Beisert/Zefa Picture Library; 86*c&b* Roland & Sabrina Michaud/The John Hillelson Agency; 86*t* John Hatt/Hutchison Library; 87*b* Roland & Sabrina Michaud/The John Hillelson Agency; 87*t* Sassoon/Robert Harding Picture Library; 88/89 Raghubir Singh/The John Hillelson Agency; 90 Victoria & Albert Museum/ The Bridgeman Art Library; 90/91 Raghubir Singh/The John Hillelson Agency; 91 Raghubir Singh/The John Hillelson Agency; 92/93 David Paterson; 94 Popperfoto; 95 David Paterson; 96/97 Tony Allen/Oxford Scientific Films; 98/99 Masahiro Iijima/ Ardea; 100/101 Dieter & Mary Plage/ Bruce Coleman; 102*b* Mary Evans Picture Library; 102*t* The Mansell Collection; 103 Dieter & Mary Plage/ Survival Anglia; 104/105 Hiroji Kubota/ Magnum/The John Hillelson Agency; 106/107 Peter Carmichael/Aspect Picture Library; 107 Butt Chak-Yu/The Stock House; 108/109 J. Bunbury Richardson/Daily Telegraph Colour Library; 110/111 Robin Morrison/ Departures; 111*b* John Topham Picture Library; 111*t* J. Allan Cash; 112/

113 Robert A. Isaacs/Photo Researchers Inc; 114/115 Tony Stone Associates; 115 Popperfoto; 116/117 Valerie Taylor/Ardea; 118 Topham Picture Library; 118/119 M. Timothy O'Keefe/Bruce Coleman; 120/121 Roger Mear/John Noble; 122/123 Kim Naylor/Aspect Picture Library; 123 Popperfoto; 124/125 Shostal Associates; 127 Robert Harding Picture Library; 128/129 Otto Rogge/NHPA; 130 Ian Griffiths/Robert Harding Picture Library; 131*b* J. Fennell/Bruce Coleman; 131*t* Ian Griffiths/Robert Harding Picture Library; 132/133 Nicholas Devore/Bruce Coleman; 134 Paul van Riel/Robert Harding Picture Library; 135 Aspect Picture Library; 136/137 Georg Gerster/The John Hillelson Agency; 138 Kevin Schafer/ Tom Stack & Associates; 139*b* Popperfoto; 139*t* Georg Gerster/The John Hillelson Agency; 140/141 Shostal Associates; 142*b* J. Allan Cash; 143 J. Allan Cash; 142*t* Guido Alberto Rossi/ The Image Bank; 144/145 Francois Gohier/Ardea; 146/147 Francois Gohier/Ardea; 147*b* Popperfoto; 147*t* L. L. T. Rhodes/Daily Telegraph Colour Library; 148/149 Shostal Associates; 150/151 Neyla Freeman/Bruce Coleman; 151 Jeff Foott/Bruce Coleman; 152/153 Karl Kummels/ Shostal Associates; 154 Neyla Freeman/Bruce Coleman; 154/155 Shostal Associates; 155 Walter Rawlings/ Robert Harding Picture Library; 156/ 157 Hiram L. Parent/Photo Researchers Inc; 158/159 Michael Freeman; 159 E. Hummel/Zefa Picture Library; 160/161 Francois Gohier/ Ardea; 162/163 J. A. L. Cooke/Oxford Scientific Films; 163 Bob McKeever/ Tom Stack & Associates; 164/165 M. P. L. Fogden/Oxford Scientific Films; 166

John Shaw/Bruce Coleman; 166/167 John Mason/Ardea; 167 Francois Gohier/Ardea; 168/169 Lucian Niemeyer; 170 John Shaw/NHPA; 170/ 171 Leonard Lee Rue III/Bruce Coleman; 172/173 Shostal Associates; 174 Michael Klinec/Bruce Coleman; 174/175 J. Allan Cash; 176/177 Tony Morrison; 178*b* Shostal Associates; 178*t* Hutchison Library; 179 Georg Gerster/The John Hillelson Agency; 180/181 Jack Jackson/Robert Harding Picture Library; 182/183 J. Allan Cash; 183 Iain Roy/David Paterson Library; 184/185 Victor Englebert/Susan Griggs Agency; 186/187 Tony Morrison; 188/189 Rosenfeld/Zefa Picture Library; 190*b* GDT-Silvestris/ NHPA; 190*t* Francisco Erize/Bruce Coleman; 191 Francisco Erize/Bruce Coleman; 192/193 P. Vauthey/Sygma/ The John Hillelson Agency; 194/195*b* Brian Hawkes/NHPA; 194/195*t* P. Vauthey/Sygma/The John Hillelson Agency; 195 P. Vauthey/Sygma/The John Hillelson Agency; 196/197 Martyn F. Chillmaid/Oxford Scientific Films; 198/199*b* Horst Munzig/Susan Griggs Agency; 198/199*t* Ian Yeomans/Susan Griggs Agency; 199 Visionbank; 200/ 201 Adrian Warren/Ardea; 202 Juan A. Fernandez/Bruce Coleman; 203*b* David Fox/Oxford Scientific Films; 203*t* Tor Eigeland/Susan Griggs Agency; 204/205 David Paterson; 206 John Cleare/Mountain Camera; 207*b* Derek Bayes/Aspect Picture Library; 207*t* The Edinburgh Photographic Library; 208/209 Owen Drayton/Bruce Coleman; 210 J. Allan Cash; 210/211 The Edinburgh Photographic Library; 211 Heather Angel; 212/213 Heather Angel; 214*b* Visionbank; 214*t* The Mansell Collection; 215 Nick Barrington/Barton Photography.